SpringerBriefs in Applied Sciences and Technology

For further volumes:
http://www.springer.com/series/8884

Leonid B. Datsevich

Conventional Three-Phase Fixed-Bed Technologies

Analysis and Critique

 Springer

Leonid B. Datsevich
University of Bayreuth
Bayreuth
Germany

and

MPCP GmbH
Bayreuth
Germany

ISSN 2191-530X   ISSN 2191-5318   (electronic)
ISBN 978-1-4614-4835-8   ISBN 978-1-4614-4836-5   (eBook)
DOI 10.1007/978-1-4614-4836-5
Springer New York Heidelberg Dordrecht London

Library of Congress Control Number: 2012943362

Printed on acid-free paper

Springer is part of Springer Science+Business Media (www.springer.com)

Preface

Three reasons inspired the author to write this monograph. The first reason is the obvious drawbacks inherent in the available literature devoted to industrial multiphase fixed-bed technologies. As a rule, the interactions between the physical and chemical phenomena are not comprehensively analyzed although an experienced reader can find a great number of publications where specific features of industrial reactors related to hydrodynamics, kinetics, mass, and heat transfer are described in detail.

Additionally, existing books suffer from a lack of the genesis of the technical solutions and, therefore, cannot pronounce the new technological perspectives that may motivate the development of more advanced techniques. Maybe, the absence of such narration in these books accounts for the reign of the technological paradigms applied to the conventional fixed-bed processes. At least for about 80 years, there have been no significant attempts to reconsider the old concepts formulated by the previous generation of process developers.

The second reason to write the monograph partly followed from the first one. Being involved in the development of new multiphase processes and discovery of unknown catalytic phenomena, the author could not understand why some professionals engaged in academic research, industrial development and operation rejected to conceive the ideas that, in the author's opinion, were not so complicated and had been proved not only in experimental units, but also in industrial installations. The explanation of such a rather strange fact lies apparently in that the traditional approaches are regarded as the undoubted postulates, even if under a close view, they turn out to be false. The monograph explains why these deeply rooted, but wrong approaches will be ineffective and even dangerous if they are realized in industrial plants. Thus, one of the objectives of this book is to help specialists to comprehend the misconceptions in scientific and design approaches as well as to indicate that some attempts to enhance industrial processes are dead-end and futureless—with no chance to succeed when these efforts follow the stereotypes embedded in the traditional literature. Additionally, the focal point of this work is to show why the conventional paradigms do not correspond to the current level of scientific and technical knowledge.

Finally, the third purpose of writing this book is to summarize the latest achievements related to new multiphase technologies and theories. Since some of these technologies were of dual use, their technical solutions in design could not be presented in detail earlier. The same restrictions were partly imposed upon the disclosure of the scientific fundamentals lying in their base. Although the description of some parts of these theories and technological solutions can be found in professional journals, their full extent is presented in this work for the first time. The author hopes that the presentation of this first-hand knowledge about the GIPKh and POLF technologies, Two-Zone Model, Oscillation theory, energy dissipation mapping and analysis of runaways as well as the lessons learnt from the revision of the old techniques can initiate fruitful discussions among professionals and motivate young researchers and practical engineers to go beyond the traditional paradigms.

The author points out that it is impossible to cover all aspects of multiphase fixed-bed reactors in one book. Therefore, the narration is concentrated on the most significant features important for the critical revision of the conventional processes with respect to intensification efforts, efficiency of the energy utilization, bottlenecks inherent in the industrial design, and process safety, regardless of the applications of new types of catalysts (monolith, regular packing, etc.). The author also tried not to overburden the text with the detailed technical specifications, equations, correlations and references, which were not necessary for the understanding of the crucial issues.

In many respects, the idea to write this book was induced by the author's opponents involved in the process development, design, operation, and the manufacture of equipment as well as the author's co-workers and Ph.D. students. The author expresses his deep appreciation to all of these people for the arguments born during their battles, many of which can be found in this work.

The author is greatly indebted to his colleagues from RSC "Applied Chemistry" (St. -Petersburg, Russia), the University of Bayreuth and MPCP GmbH (Bayreuth, Germany) for the support and discussions. The author would like to mention some of them here: H. M. Avanesova, O. D. Ignatieva, M. P. Kambur, G. A. Mironova, D. A. Mukhortov, M. I. Nagrodskii, Y. V. Sharikov, O. J. Sokolova, N. G. Zubritskaya, B. Battsengel, A. Jess, S. Fritz, T. Oehmichen, C. Schmitz, W. Wache, S. Werth, P. Dallakian, F. Grosh, R. Wolfrum.

Contents

List of Symbols

a_s	Specific external area of catalyst particles in the bed (external surface area of catalyst particles per bed volume), m^{-1}
$C_{A,0}$	Concentration of the liquid reactant in the liquid feed, mol/m^3
$C_{A,cat}$	Concentration of a liquid reactant in the catalyst bulk, mol/m^3
$C_{A,in}$	Concentration of a liquid reactant at the reactor inlet in the GIPKH and POLF technologies (after mixing with the recirculated product), mol/m^3
$C_{A,inv}$	Concentration of a liquid reactant according to Eq. (5.19), mol/m^3
$C_{A,l}$	Concentration of a liquid reactant in the liquid bulk, mol/m^3
$C_{A,out}$	Concentration of the liquid reactant at the outlet of the reactor or catalyst bed, mol/m^3
$C_{A,out}^{(1)}$	Concentration of the liquid reactant at the outlet of the first catalyst bed in the two-stage POLF technology (Fig. 9.4b), mol/m^3
$C_{A,out}^{(2)}$	Concentration of the liquid reactant at the outlet of the second catalyst bed in the two-stage POLF technology (Fig. 9.4b), mol/m^3
$C_{A,s}$	Concentration of a liquid reactant on the external catalyst surface, mol/m^3
ΔC_A	Concentration drop of the liquid reactant $\Delta C_A = \frac{1}{n}\left(\frac{C_{B,g}}{H}\right)$, mol/m_3
$C_{B,cat}$	Concentration of a gas compound in the catalyst bulk, mol/m^3
$C_{B,g}$	Concentration of a gas compound in the gas phase, mol/m^3
$C_{B,g}^*$	Equilibrium concentration of a gas compound on the gas side of the gas–liquid interface, mol/m^3
$C_{B,l}$	Concentration of a gas compound in the liquid bulk, mol/m^3
$C_{B,l}^*$	Equilibrium concentration of a gas compound on the liquid side of the gas–liquid interface, mol/m^3
$C_{B,out}$	Concentration of the gas reactant at the outlet of the reactor in the one-stage POLF process (Fig. 9.4a), mol/m^3
$C_{B,s}$	Concentration of a gas compound on the external catalyst surface (in the liquid phase), mol/m^3
$C_{P,g}$	Molar heat capacity of the gas phase for constant pressure, $J/(mol\ K)$

$C_{P,l}$	Specific heat of the liquid phase for constant pressure, J/(kg K)
$C_{V,g}$	Specific heat of the gas phase for constant volume, J/(mol K)
d_{cat}	Diameter of a catalyst particle, m
d_{HE}	Tube diameter of an incorporated heat exchanger (Appendix A), m
d_{tube}	Tube diameter, m
$D_{A,l}$	Diffusivity of a liquid compound in the liquid phase, m^2/s
$D_{B,eff}$	Effective diffusion coefficient of gas inside the catalyst pellet, m^2/s
$D_{B,l}$	Diffusivity of a gaseous compound in the liquid phase, m^2/s
D_r	Diameter of reactor, m
E	Power demanded for gas compressing or power lost at gas transportation, W
$E_{recycle}$	Estimated power demanded for gas recycling by the recycle compressor (Table 3.2), W
F_{HE}	Heat transfer surface of an incorporated heat exchanger (Appendix A), m^2
$F_{reactor}$	Cross-sectional surface of the catalyst bed, m^2
g	Acceleration of gravity, $g = 9.81$ m/s^2
G.S.	Surplus of the gas compound over the liquid reactant (Eq. (5.16))
H	Henry's coefficient $(C_{B,g}^* = HC_{B,l}^*)$
$-\Delta H_A$	Reaction heat related to liquid compound $(-\Delta H_A = -n\Delta H_B)$, J/mol
$-\Delta H_B$	Reaction heat related to gas compound $(-\Delta H_B = -\Delta H_A/n)$, J/mol
j_A	Molar flux of the liquid reactant, mol/(m^2 × s)
j_B	Molar flux of the reacting gas, mol/(m^2 × s)
k_V	Intrinsic first-order reaction-rate constant per unit volume of a catalyst pellet, s^{-1}
K_{g-l}	Total gas–liquid mass transfer coefficient of a gas reactant, m/s
K_{g-s}	Overall gas–liquid–solid mass transfer coefficient, m/s
$K_{recycle}$	Recirculation rate of the liquid phase
l_{bed}	Total length of a catalyst bed, m
l_{cat}	Coordinate of the catalyst length in the flow direction, m
l_{inv}	Coordinate of the inversion point, at which the mass transfer limiting stage changes, m
L_{HE}	Total length of tubes of an incorporated heat exchanger (Appendix A), m
L_{tube}	Tube length, m
L.S.	Surplus of the liquid reactant over the gas compound at the reactor inlet (see Eq. (5.15))
n	Stoichiometric coefficient in Eq. (3.1)
n_1	Index of power in Eq. (5.13)
n_2	Index of power in Eq. (5.13)
n_3	Index of power in Eq. (5.14)
$N_{A,feed}$	Molar flow rate of liquid reactant feed, mol/s
$N_{B,feed}$	Molar flow rate of gas feed, mol/s
$N_{g,recycle}$	Molar flow rate of recycled gas, mol/s

P	Pressure, N/m^2 (or bar)
P^*	Arithmetical mean of the inlet and outlet pressures at the tube ends, N/m^2 (or bar)
P_B	Partial pressure of a reacting gas, N/m^2 (or bar)
P_1	Pressure at the compressor intake or somewhere in the gas loop, N/m^2 (or bar)
ΔP	Pressure drop over the chosen element of the gas loop, N/m^2 (or bar)
ΔP_Σ	Pressure difference between the outlet and intake of a recycle compressor, N/m^2 (or bar)
$Q_{l,\text{feed}}$	Volumetric flow rate of liquid feed, m^3/s
$Q_{l,\text{feed}}^{\text{peak}}$	Peak flow rate of liquid at feed modulation, m^3/s
Q_{HE}	Heat removed by incorporated heat exchanger, W
Q_{recycle}	Volumetric flow rate of the circulating liquid, m^3/s
r_{true}	Maximum (intrinsic) reaction rate related to a mole of a liquid reactant per volume of the catalyst particle if there were no concentration gradients inside, mol/(m^3 × s)
r_V	Overall (observed) reaction rate related to a mole of a liquid reactant per volume of the catalyst particle, mol/(m^3 × s)
R	Universal gas constant, $R = 8.314$ J/(mol × K)
Re	Reynolds number
S.P.	Specific productivity of the catalyst bed, s^{-1}
T	Temperature, K
T_0	Temperature of the gas–liquid or liquid mixture at the reactor inlet, K
T_1	Temperature at the compressor intake, K
T_{max}	Maximum allowable temperature preconditioned by the process selectivity, catalyst aging and safety, K
T_{min}	Minimal process temperature preconditioned by the appropriate reaction rate, K
ΔT	Adiabatic temperature rise of the reaction mixture, K
ΔT_{ad}	Maximum adiabatic temperature rise of the reaction mixture in the absence of heat removal, K
ΔT_{HE}	Log mean temperature difference in a heat exchanger, K
ΔT_{permit}	Permitted (adiabatic) temperature rise of the reaction mixture, K
U_g	Superficial velocity of gas equal to volumetric flow rate per unit cross-sectional area of packed bed, m/s
U_l	Superficial velocity of liquid equal to volumetric flow rate per unit cross-sectional area of packed bed, m/s
U_{tube}	Velocity of the gas phase in the tube, m/s
V	Bulk volume of catalyst in the bed, m^3
V_{bed}	Catalyst bed volume, m^3
V_{HE}	Volume of heat exchanger tubes (Appendix A), m^3
X	Conversion of the liquid reactant ($X = (1 - C_{A,\text{out}}/C_{A,0})$)
X_B	Conversion of the gas reactant in the one-stage POLF process (Fig. 9.4a) ($X = (1 - C_{B,\text{out}}/C_{B,l}^*)$)

Greek Symbols

α	Heat transfer coefficient, $W/(m^2 \times K)$
$\beta_{A,s}$	Liquid–solid mass transfer coefficient of a liquid reactant, m/s
$\beta_{B,g}$	Gas-side mass transfer coefficient of a gas reactant, m/s
$\beta_{B,l}$	Liquid-side mass transfer coefficient of a gas reactant, m/s
$\beta_{B,s}$	Liquid–solid mass transfer coefficient of a gas reactant, m/s
γ	Adiabatic index of gas $\gamma = C_{P,g}/C_{V,g}$
δ	Film thickness, m
ε	Bed porosity
$\zeta(Re)$	friction factor
η	Effectiveness factor $\eta = r_V/r_{true}$
η_g	Viscosity of the gas phase, $N{\cdot}s/m^2$
η_l	Viscosity of the liquid phase, $N{\cdot}s/m^2$
μ_g	Molar mass of gas, kg/mol
ρ_g	Density of the gas phase, kg/m^3
ρ_l	Density of the liquid phase, kg/m^3
τ_1	Feed time in unsteady-state operation mode, s
τ_Σ	Period of modulation, s
φ	Thiele module
ψ	Split

Dimensionless Groups

Re_g	Reynolds number for gas flow based on U_g and d_{cat}
Re_l	Reynolds number for liquid flow based on U_l and d_{cat}
$Sc_{A,l}$	Schmidt number for a liquid reactant
$Sc_{B,l}$	Schmidt number for a gas compound in liquid
$Sh_{i,s}$	Sherwood number for liquid–solid mass transfer, liquid compound $i = A$, gas compound $i = B$
$Sh_{B,l}$	Sherwood number for mass transfer (liquid-side) of a gas compound in liquid

Subscripts

A	Liquid reactant
B	Gaseous reactant
cat	Catalyst
g	Gas phase

i Gas or liquid compound ($i = A$ or $i = B$, respectively)
l Liquid phase

Acronyms

BCR   Bubble column reactor
CSTR   Continuously operated stirred-tank reactor
GIPKh   Russian abbreviation of The State Institute of Applied Chemistry
MTR   Multitubular reactor
PFR   Plug flow reactor
POLF   Presaturated one-liquid flow
TBR   Trickle-bed reactor

Chapter 1
Introduction

Three-phase fixed-bed reactors are very often encountered in industrial applications for carrying out different chemical reactions between gaseous and liquid reactants on porous catalysts, in processes such as, hydrogenation, hydrotreating, purification, the Fischer–Tropsch synthesis, and in many others. These processes form the basis for production of a large variety of intermediate and ultimate products in refinery, bulk and fine chemistry, in manufacture of monomers, solvents, pharmaceuticals, fragrances, fuels, food additives, etc.

There are three conventional types of fixed-bed reactors, which are mainly employed for multiphase processes: Trickle-Bed Reactors (TBRs), Bubble (packed) Column Reactors (BCRs), and MultiTubular Reactors (MTRs).

Some typical reactions realized in these reactors are exemplified in Table 1.1. It is worth pointing out that it is impossible to enumerate all industrial applications in one table so that each example listed in Table 1.1 symbolizes a great number of particular reactions occurring in the industry. Furthermore, several types of reactions given in Table 1.1 can be accomplished at the same time in the same reactor since some compounds can have different functional groups, which react in the course of consecutive or parallel reactions (e.g. hydrogenation of nitro compounds, oil hydrotreating, etc.).

As is known, the realization of each heterogeneous, multiphase industrial process is always a difficult compromise between the productivity, selectivity, safety, lifetime of a catalyst and its accessibility, complexity of technical embodiments, and other process features, the balance among all of which should result in reasonable production and investment costs (see Fig. 1.1).

In the following consideration, we will show that conventional fixed-bed reactors do not correspond to the current demand from the points of view of efficiency, technical configuration, and process safety.

Moreover, as will be shown, these reactors do not possess any potential for development. Any attempt to improve substantially the reaction performance, e.g., to enhance the reactor productivity (not by several percents as it is always

L. B. Datsevich, *Conventional Three-Phase Fixed-Bed Technologies,*
SpringerBriefs in Applied Sciences and Technology,
DOI: 10.1007/978-1-4614-4836-5_1, © The Author(s) 2012

Table 1.1 Typical reactions realized in industrial fixed-bed reactors

N	Reaction	Schematic reaction equation	Conditions	Reaction heat related to converted gas $(-\Delta H_B)$ (kJ/mol$_B$)
1	Hydrogenation of aldehydes to alcohols [1, 2]	$R - CHO + H_2 \rightarrow R - CH_2OH$	50–150 °C 30–100 bar	67–83
2	Hydrogenation of ketones to alcohols [2–4]	$R_1\text{-}\overset{O}{C}\text{-}R_2 + H_2 \rightarrow R_1\text{-}\overset{OH}{CH}\text{-}R_2$	50–150 °C 30–100 bar	55–60
3	Hydrogenation of double and triple bonds (e.g. in alkenes, fatty acids, alkynes, etc.) [2–4]	$R_1 = R_2 + H_2 \rightarrow R_1H - R_2H$ $R_1 \equiv R_2 + H_2 \rightarrow R_1H = R_2H$ $R_1 = R_2 + 2H_2 \rightarrow R_1H_2 - R_2H_2$	80–200 °C 150–300 bar	113–156
4	Hydrogenation of acids to alcohols [1, 2]	$RCOOH + 2H_2 \rightarrow RCH_2OH + H_2O$	130–250 °C 100–300 bar	19–21
5	Hydrogenation of esters to alcohols [5, 6]	$R_1\text{-}\overset{O}{C}\text{-}O\text{-}R_1 + 2H_2 \rightarrow R_1\text{-}CH_2OH + R_2OH$	170–250 °C 150–300 bar	38–50
6	Hydrogenation of nitriles to primary amines [1, 7]	$RC \equiv N + 2H_2 \rightarrow RCH_2 - NH_2$	50–150 °C 20–100 bar	67–80
7	Hydrogenation of nitroso compounds to amines	$R - NO + 2H_2 \rightarrow R - NH_2 + H_2O$	80–150 °C 200 bar	130–182
8	Hydrogenation of nitrosamines to hydrazines	$RN - NO + 2H_2 \rightarrow RN - NH_2 + H_2O$	40–100 °C 50–200 bar	130–182
9	Hydrogenation of nitrocompounds to amines [2]	$R - NO_2 + 3H_2 \rightarrow R - NH_2 + 2H_2O$	70–150 °C 50–200 bar	146–182
10	Hydrogenation of aromatic nitrocompounds to aromatic amines [7]		90–150 °C 50–150 bar	182
11				182

(continued)

Hydrogenation of chlornitro compounds to chloramines [7]	(see structure)	70–150 °C 100–150 bar	
12	Purification processes (elimination of unsaturated, oxygen and halogen containing compounds by hydrogenation)	Nearly all reactions enumerated in this table	50–120 °C 10–70 bar	
13	Hydrodehalogenation [7]	$RCl + H_2 \rightarrow RH + HCl$	70–150 °C 100–150 bar	65
14	Ring hydrogenation [2–4, 7]	(see structure)	150–230 °C ~100 bar	63–69
15	The Fischer–Tropsch synthesis [7–9]	$CO + 2H_2 \rightarrow -CH_2- + H_2O$	180–250 °C 10–45 bar	165
16	Hydrotreating [10][a]		320–430 °C 30–70 bar	
16.1	Hydrodesulfurization of mercaptans, sulfides and disulfides	$RSH + H_2 \rightarrow RH + H_2S$ $R_1SR_2 + H_2 \rightarrow R_1H + R_2H + H_2S$ $R_1-S-S-R_2 + 3H_2 \rightarrow R_1H + R_2H + 2H_2S$		58–70 52 44
16.2	Hydrodesulfurization of thiophene	(see structure) $+4H_2 \rightarrow C_4H_{10} + H_2S$		65
16.3	Hydrodesulfurization of dibenzothiophenes	(see structure)		63–69
16.4	Hydrodenitration	(see structure) $+5H_2 \rightarrow C_5H_{12} + NH_3$		78
16.5	Deoxigenation			28

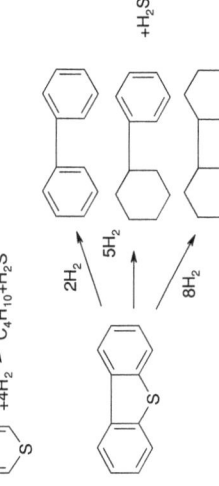

(continued)

$$\text{OH} \quad \xrightarrow{+H_2} \quad + H_2O$$

17   Hydrocracking [3, 7]

$$R_1 - R_2 + H_2 \rightarrow R_1H + R_2H$$

375–430 °C   42–65
100–170 bar

[a] Since any industrial hydrotreating process encompasses a great number of different reactions, which include not only those enumerated in N. 16, but also hydrocracking and hydrogenation of other unsaturated compounds, the reaction heat of hydrotreating depends on the refinery stream to be processed. For the purpose of a reactor design, two generalized characteristics with regard to heat production are usually used. The first one is the heat effect, which is related to reacting hydrogen. This value is usually taken about $60 \div 80$ kJ per mole of H_2 consumed. The second important parameter is the chemical consumption of hydrogen per volume of a treated stream, for instance, $2 \div 10$ Nm^3_{H2}/m^3_{oil} for naphtha or $102 \div 450$ Nm^3_{H2}/m^3_{oil} for heavier fractions and pyrolytic oils (see, for example, [11–13]). Proceeding from these two generalized parameters, the heat effect related to 1 m^3 of hydrotreated oil can be estimated equal to $5.4 \div 1610$ MJ/m^3_{oil}

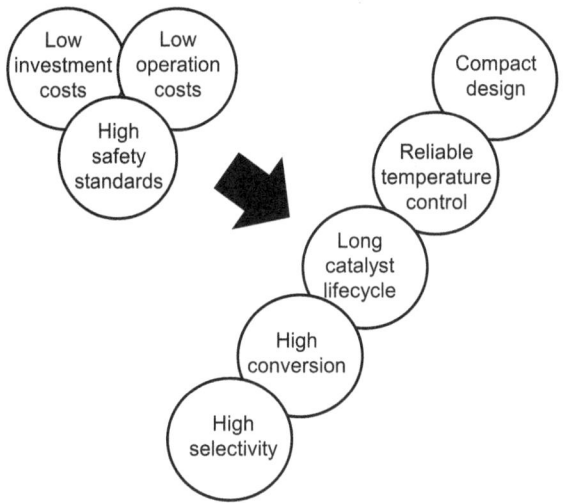

Fig. 1.1 Request for design of industrial units

Fig. 1.2 Vehicle driven by a horse: Is there a potential for improvement if the horse is used as a driving force?

discussed in the available literature, but by several times) cannot be realized even at the expense of an over-proportional increase in the energy input or pressure because of the physical and chemical restrictions inherent in these technologies.

The author would like to forestall any accusations, which can be brought against him, if someone advances an argument with regard to some improvements that have really been demonstrated by the introduction of new catalysts, by more effective

peripheral equipment (compressors, heat exchangers, etc.), and by more effective
heat management (recuperation, heating, cooling, etc.). The author would like only
to emphasize the cosmetic nature of such achievements although they can be
essential for a specific industrial process.

Hence, let us define the absence of "potential for development" in such a
meaning as is explained in Fig. 1.2. From the current point of view, will anybody
regard the horse-drawn car as a device possessing any "potential for development
(to drive faster, effectively, etc.)" even if it is furnished with the most modern
equipment? For example, will anybody seriously consider the fact of mounting the
cruise or electronic stability control in this car as an action toward the substantial
development if the horse remains to be the driving force?

References

1. P. Rylander, *Catalytic hydrogenation in organic synthesis* (Academic Press, New York, 1979)
2. N.N. Lebedev, Chemistry and technology of basic organic and petrochemical synthesis (in Russian) (Moscow, Khimija, 1981)
3. S.B. Jaffe, Kinetics of heat release in petroleum hydrogenation. Ind. Eng. Chem. Process. Des. Dev. **13**(1), 34–39 (1974)
4. D.W. Rogers, *Heats of hydrogenation* (World Scientific, New Jersey, 2006)
5. E.V.W Gritz, in Handbook of Heterogeneous Catalysis, ed. by G. Ertl, H. Knözinger, J. Weitkamp. *Fat Hydrogenation*, vol. 5 (Wiley-VCH, Weinheim, 1997), pp. 2224–2227
6. G. Darsow, G.M. Petruck, H.J. Alpers, European Patent Method for the preparation of aliphatic alpha, omega-*diols*, 0,72,1928, 1996
7. C.H. Bartholomew, R.J. Farrauto, *Fundamentals of industrial catalytic processes* (Wiley, New Jersey, 2006)
8. B. Jager, Development of commercial Fischer Tropsch reactors, in *AIChE 2003 Annual Meeting Conference Proceedings* (AIChE, New Orleans, 2003), http://www.fischer-tropsch.org/primary_documents/presentations/AIChE%202003%20Spring%20National%20Meeting/BJager-DvlpFTReactor.pdf
9. V.I. Anikeev, A. Yermakova, B.L. Moroz, in The state of studies of the Fischer-Tropsch srocess in Russia, *AIChE 2003 Spring National Meeting* (New Orleans, 2003) http://www.fischer-tropsch.org/primary_documents/presentations/AIChE%202003%20Spring%20National%20Meeting/Presentation%2080c%20Russia.pdf
10. T. Kabe, A. Ishihara, W. Quian, *Hydrodesulfurization and Hydrodenitrogenation* (Wiley-VCH, Weinheim, 1999)
11. A.G. Bridge, in Handbook of Petroleum Refining Processes, ed. by R.A. Meyers. *Hydrogen Processing* (McGraw-Hill, New York, 1996), pp. 14.1–14.68
12. P.R. Robinson, G.E. Dolbear, in Practical Advances in Petroleum Processing, ed. by C.S. Hsu, P.R. Robinson. *Hydrotreating and Hydrocracking: Fundamentals*, vol. 1 (Springer, New York, 2006), pp. 177–218
13. J. Wildschut, F.H. Mahfud, R.H. Venderbosch, H.J. Heeres, Hydrotreatment of fast pyrolysis oil using heterogeneous noble-metal catalysts. Ind. Eng. Chem. Res. **48**(23), 10324–10334 (2009)

Chapter 2
Technological Reasons for the Selection of Fixed-Bed Reactors

The choice for an appropriate reactor for carrying out a multiphase reaction on a solid catalyst is always a rather challenging task. Not going deeply into the particularities inherent in a specific process, let us point out three main problems that should be solved in all multiphase reactions mentioned in Table 1.1 regardless of their chemical nature.

The first problem is a request for sufficient mass transfer of all reacting compounds to the catalyst surface, especially if an active catalyst is employed. The second point is a demand on the removal of heat generated in the course of the reaction, and the third problem is associated with striving for the full utilization of the catalyst potential.

At first look, the reactors with a suspended catalyst are more favorable for multiphase reactions.

Actually, due to violent stirring, it is easy to arrange the necessary intensity of gas–liquid and liquid–solid mass transfer and to provide the heat removal by installation of a heat exchanger inside the slurry reactor or outside it (e.g. in combination with an external liquid loop).

Fine catalyst particles nearly always allow one to utilize the whole bulk volume of a single catalyst particle so that the internal surface becomes completely accessible for reactants (no concentration gradients), which ensures a high reaction rate.

However, slurry reactors are reluctantly applied in the industry, especially in the medium- and large-scale production because of two paramount problems.

The first is caused by the necessity to separate a catalyst from a product. Even if relatively big particles are initially charged into the reactor, they—being subjected to intensive stirring—get less and less in the course of the process. Finely powdered particles always worsen the ability of any filter to perform the filtration by blocking the filtering surface or passing through the filter texture. Multistage filtration is often needed to exclude catalyst penetration into the product. Moreover, if sticky, high-molecular compounds are formed, the filtration becomes

L. B. Datsevich, *Conventional Three-Phase Fixed-Bed Technologies*,
SpringerBriefs in Applied Sciences and Technology,
DOI: 10.1007/978-1-4614-4836-5_2, © The Author(s) 2012

Table 2.1 Slurry reactors versus conventional fixed-bed reactors

N	Compared features	Slurry reactor	Fixed-bed reactors
1	Reaction performance		
1.1	Overall reaction rate related to the catalyst mass	High	Low
1.2	Overall reaction rate related to the reactor volume (or volume of all apparatuses)	Low	High
1.3	Pressure	Low/Medium	Medium/High
2	Safety		
2.1	Temperature control	Simple	Simple
2.2	Probability of runaways in exothermic reactions	Low	Extremely high
2.3	Special measures against possible runaways in exothermic reactions	Not necessary	Nearly always necessary
3	Economic efficiency		
3.1	Operating costs	High	Low
3.2	Investment costs	High (at continuous operation)	Low
3.3	Operation mode	Discontinuous (continuous operation is very seldom)	Continuous
3.4	Number of apparatuses in the continuous operation	A lot	A few
3.5	Filtration stage	Necessary	Absent
3.6	Demand on production flour area/working space	Great	Low
3.7	Multi-product operation in the case of medium scale output	More cost based	Less cost based
3.8	Catalyst consumption	High	Low

by far more dramatic by virtue of a gluing filter cake so that a frequent replacement of filters is needed.

The second problem is related to the abrasive properties of many commercial catalysts. Due to the attrition of stirrers, tubes, valves, control devices, etc., much time is demanded for the revision and replacement of apparatuses, which results in a rise in operating costs.

If a slurry reactor should operate in continuous mode, more difficulties should be solved additionally. For example, they encompass the catalyst preparation (e.g. activation), continuous charge, discharge and deactivation of a pyrophoric catalyst, replacement or regeneration of a filter, and so on.

It is not surprising that continuous slurry reactors are very seldom encountered in industrial applications if a comparatively large output is needed.

The only method that makes it possible to evade the problems attributed to a suspended catalyst is the fixation of catalyst pellets inside a reactor (column or tube). In such fixed-bed reactors, the conventional way to achieve the appropriate mass transfer and simultaneously to remove reaction heat is the passage of a reacting gas through the catalyst bed. Since the chemical consumption of the reacting gas is much less than the amount of the gas which should go through the reactor, the unreacted part of the gas is mixed with a fresh portion and returned back to the reactor by means of the gas recirculation. Such reactors with gas loop gain widespread acceptance in different industries.

Unfortunately, it is impossible to utilize the whole catalyst potential in industrial processes since small catalyst particles cannot be used in fixed-bed reactors. Actually, it is impracticable to apply particles less than 1 mm in size because of the pressure drop problem (Sect. 5.2). Despite the fact that fixed-bed reactors employ catalysts of 1–10 mm, which inevitably leads to the existence of intraparticle diffusion limitations, the fixed-bed technology considerably exceeds the efficiency of the slurry technique with regard to many practical features (see Table 2.1). For instance, it is impossible to imagine any refinery hydrotreating process of throughput about 100 m^3 of oil per h carried out in a slurry reactor.

Chapter 3
Traditional Approaches to the Design of Fixed-Bed Reactors

For the purpose of the successful analysis of existing technologies, we will try to reproduce the main arguments that were likely made by the previous generations of process developers. On the one hand, it is interesting to retrace why the original ideas have established such a rigid paradigm that is widely shared by the industrial and scientific communities and, by force of habit, remains without any critical view up to now. On the other hand, it becomes clear why the process layouts in many applications look so uniform even when it is not demanded by the physical or chemical nature [e.g., purification processes, including ultra-deep hydrodesulfurization (see Chap. 6)].

The typical reaction scheme in multiphase processes can be expressed as

$$A + nB \rightarrow Product \text{ (liquid or gas)} \tag{3.1}$$

where A means an initial liquid compound to be converted, for instance, nitrobenzene, B is a reacting gas (e.g., hydrogen), *Product* represents one or more liquid or gas products, e.g., aniline and water, and n is the stoichiometric coefficient, which is equal, for example, to three in the case of complete nitrobenzene hydrogenation to aniline.

Since the reactions between hydrogen and some organic species, which are liquid or residing inside the liquid phase, represent the majority of industrial gas–liquid–solid processes, in the following consideration, we will purposefully focus on them.

It is worth pointing out that hydrogenation according to Eq. (3.1) makes no differences toward many other industrial reactions, all of which have a great number of common features including low gas solubility, high reaction heat, similar process embodiments, and so on.

L. B. Datsevich, *Conventional Three-Phase Fixed-Bed Technologies*,
SpringerBriefs in Applied Sciences and Technology,
DOI: 10.1007/978-1-4614-4836-5_3, © The Author(s) 2012

3.1 Approaches from the Point of View of Kinetics

It is traditionally assumed that the porous structure of an individual catalyst particle is completely occupied with liquid due to strong capillary effects (e.g. [1]). It is also supposed that the reactor design should ensure the possibly even liquid and gas distribution in any cross section, so that all particles in the radial direction of an industrial reactor have uniform concentrations of reacting compounds on the catalyst surface as well as the same temperature. (Note: In MTRs, there can be observed a gradient of temperature and concentration in the radial direction, which should be taken into consideration at the scaling-up and analysis of process safety.)

The reaction rate r_V related to the volume of a single catalyst pellet, which is situated somewhere in the catalyst bed, can be determined by a function including such parameters as catalyst size d_{cat}, temperature T and concentrations of gas and liquid compounds on the liquid–solid interface ($C_{B,s}$ and $C_{A,s}$) and can be written as

$$r_V = f(d_{cat}, T, C_{B,s}, C_{A,s}) \tag{3.2}$$

(Note: Eq. (3.2) represents the simplest case when there is no dependence of the reaction rate on the concentrations of a final product(s), intermediate species, by-products, and a solvent).

The dependence of the reaction rate on a catalyst particle size is usually interpreted with the help of the Thiele/Zeldovich model [2, 3] in terms of the effectiveness factor η as

$$r_V = \eta r_{true} \tag{3.3}$$

where r_{true} represents the maximal reaction rate, which a catalyst particle would demonstrate if there were no concentration gradients inside. For the simple kinetics, as, for example, given by equation

$$r_{true} = k_V C_{B,cat} \tag{3.4}$$

The effectiveness factor for a spherical particle with homogeneous properties can be expressed through the so-called Thiele module φ

$$\varphi = \frac{d_{cat}}{2} \sqrt{\frac{k_V}{D_{B,eff}}} \tag{3.5}$$

as

$$\eta = \frac{3}{\varphi} \left(\frac{1}{\tanh \varphi} - \frac{1}{\varphi} \right) \tag{3.6}$$

(Note: In the case of the more complex kinetic relations than given by Eq. (3.4), it may not be possible to express the effectiveness factor η analytically as given by Eq. (3.6)).

As a rule, industrial chemical engineers aiming at a certain industrial application do not carry out the detailed investigation with regard to how the pellet size influences the reaction performance. The explanation of this lies in the stringent boundary conditions inherent in any industrial production. On the one hand, it is preferable to apply as small particles as possible in order to increase the overall reaction rate. On the other hand, the smaller the catalyst particles, the higher the pressure drop built along the catalyst bed. Thus, the choice of the catalyst size is predetermined by the reasonable pressure drop caused by the passage of the reacting mixture through the reactor.

The temperature is one of the most influential process parameters. It dictates the reactor productivity, process selectivity, and catalyst lifetime. In all industrial applications, the value of the temperature along the catalyst bed must strongly be restricted. Its choice represents a difficult compromise between productivity, selectivity, and a rate of the catalyst aging. From the point of view of the compact reactor design, it is preferable to carry out the process under possibly higher temperatures in order to have a faster reaction rate (higher reactor productivity). However, the high temperature inevitably leads to a rise in an amount of by-products, which worsens the product quality. Moreover, the increased temperature accelerates the catalyst decay due to the formation of coke or sticky, high molecular and polymer-like compounds in catalyst pores.

Another aspect of high temperature with regard to kinetics is the process safety. Any final product in the multi-phase process according to Eq. (3.1) can always be regarded as an intermediate substance, which can react still further. Even if the catalyst of the best selectivity is used, under increased temperatures, any goal product can undergo the further conversion according to reactions presented in Table 1.1. By-reactions are accompanied by far more heat liberation, which, in turn, provokes further reactions with still higher heat generation.

For example, hydrogenation of furfural to furfuryl alcohol as a goal product can proceed thereafter to tetrahydrofurfuryl alcohol and still further up to hydrocarbons of lower molecular weight:

$$(3.7)$$

In the case of the worst scenario concerning by-reactions, the heat production in an industrial reactor can rise sharply by several times, which leads to a very strong and quick growth in temperature and, therefore, in pressure, resulting in reactor damage. (Note: If the reaction is carried out in the presence of an organic solvent, the high temperature can also cause its conversion with additional heat liberation).

The concentrations of the reacting gas and liquid species $C_{B,s}$ and $C_{A,s}$ on the external catalyst surface are also very crucial parameters. Usually, it is assumed that the reaction runs faster at their higher concentrations although it is correct to some extent. If these concentrations are as high as in the liquid bulk (i.e. there is no liquid–solid mass transfer resistance), the reaction rate is expected to be also high.

However, due to the mass transfer resistance in the case of a fast reaction, the surface concentration of a gas or liquid compound can be so insignificant that the reaction rate can be determined exclusively by external mass transfer of this key compound (see analysis in Sect. 5.5).

The dependence of the reaction rate on the surface concentrations of reactants as given by Eq. (3.2) is often defined through the chemical mechanism, which involves all elementary steps such as diffusion in pores, adsorption and desorption of reactants, and products on active catalyst sites, and finally the reaction itself.

The intrinsic (true) chemical reaction rate r_{true} (i.e., received in the absence of the concentration gradients inside a catalyst bulk, when, for example, an extremely crushed catalyst is tested) represents, as a rule, the complex function. This relation is usually interpreted in terms of the surface concentrations of adsorbed species with the help of the Langmuir–Hinshelwood-Hougen–Watson (LHHW) or Eley–Rideal (ER) approaches.

Unfortunately, real chemical mechanisms in the overwhelming number of industrial processes are likely to remain veiled for a long time if not forever. At least for now, it is difficult (if possible at all) experimentally to observe the real interaction of reacting molecules with the catalyst surface and between adsorbed species under real process conditions.

In order to yield the equation for an intrinsic reaction rate, various hypotheses with regard to the elementary steps, which may occur in the course of reacting and adsorbing, are proposed for modeling. Such multitudinous assumptions should consider different adsorption mechanisms (competitive or noncompetitive, with or without dissociation), and different reacting mechanisms (only adsorbed species react with each other, or one of the reactants is not absorbed being free in the liquid phase and reacts with another one that is adsorbed) [4, 5]. By cross coupling all considered elementary steps, a variety of functions for the intrinsic reaction rate with a great number of model coefficients can be produced. The number of these coefficients can become still more. Actually, if Eq. (3.1) represents a total result of a number of consecutive/parallel reactions, the intermediate species of these reactions can also be included in the intrinsic model. The set of the coefficients in the kinetic model can be established with the help of the approximation algorithm that allows one to fit the experimental data.

For example, Table 3.1 illustrates only some limited number of the functions for the intrinsic reaction rate corresponding to the simplest cases.

It is necessary to point out that such a great set of the coefficients, which can never be determined in the course of independent experiments, nearly always permit—with a relevant deviation—to match nearly each kinetic model to the experimental data by an appropriate choice of numerical values for the coefficients (for example, see [6]).

(Note: According to the author's opinion, the intrinsic kinetics yielded by such an approach can seldom be recommended for design purposes, let alone its incorporation into the Thiele/Zeldovich model and into the general, universal reactor model including heat and mass transfer phenomena, especially, under nonsteady conditions).

Table 3.1 Exemplification of the LHHW model for the reaction given by Eq. (3.1)

Supposed mechanism	Intrinsic reaction rate	Limiting case		
		$C_{A,cat} \to \infty$ $C_{B,cat} \to 0$	$C_{A,cat} \to 0$ $C_{B,cat} \to \infty$	$C_{A,cat} \to 0$ $C_{B,cat} \to 0$
Noncompetitive adsorption, without dissociation of gas	$r = \dfrac{kC_{A,cat}C_{B,cat}}{(1+k_A C_{A,cat})(1+k_B C_{B,cat})}$	$r \sim C_{B,cat}$	$r \sim C_{A,cat}$	$r \sim C_{A,cat}C_{B,cat}$
Noncompetitive adsorption, with dissociation of gas	$r = \dfrac{kC_{A,cat}C_{B,cat}^{0.5}}{(1+k_A C_{A,cat})(1+k_B C_{B,cat}^{0.5})}$	$r \sim C_{B,cat}^{0.5}$	$r \sim C_{A,cat}$	$r \sim C_{A,cat}C_{B,cat}^{0.5}$
Competitive adsorption, without dissociation of gas	$r = \dfrac{kC_{A,cat}C_{B,cat}}{(1+k_A C_{A,cat}+k_B C_{B,cat})^2}$	$r \sim \dfrac{C_{B,cat}}{C_{A,cat}}$	$r \sim \dfrac{C_{A,cat}}{C_{B,cat}}$	$r \sim C_{A,cat}C_{B,cat}$
Competitive adsorption, with dissociation of gas	$r = \dfrac{kC_{A,cat}C_{B,cat}^{0.5}}{(1+k_A C_{A,cat}+k_B C_{B,cat}^{0.5})^2}$	$r \sim \dfrac{C_{B,cat}^{0.5}}{C_{A,cat}}$	$r \sim \dfrac{C_{A,cat}}{C_{B,cat}^{0.5}}$	$r \sim C_{A,cat}C_{B,cat}^{0.5}$

Note The number of the coefficients in the equations for the intrinsic reaction rate can significantly be multiplied if, for example, (i) sorption of a solvent or product is taken into account, or (ii) intermediates are introduced in the reaction mechanism

Even if the numerical solutions hit temperature and concentration profiles received in the pilot-plant fixed-bed reactor, none can guarantee the reliability of the model for the purposes of scaling-up and control. From this point of view, the macrokinetic data in the form of Eq. (3.2) obtained in the experiments, where the catalyst particles of the real industrial size are used, are more preferable.

Nevertheless, some fundamental conclusions can be made from the kinetic mechanisms suggested, for example, in Table 3.1 with regard to the change of the reaction order. Such a change in the reaction behavior can be conditioned by altering the concentrations of reactants as they pass the reactor. As a rule, the concentrations of the reacting species on the external catalyst surface $C_{B,s}$ and $C_{A,s}$ do not remain constant along the catalyst bed length. Even if the partial pressure of the gas compound is unchanged in the gas phase throughout the fixed-bed reactor, in comparatively fast reactions, its concentration on the solid surface $C_{B,s}$ can drastically vary from minimum to maximum [7].

As will be shown in Sect. 5.5, two areas in the catalyst bed can be specified: (1) the initial part of the catalyst bed situated near to the inlet of the liquid reactant and (2) the finishing part of catalyst bed downstream where there is a considerable concentration of the product.

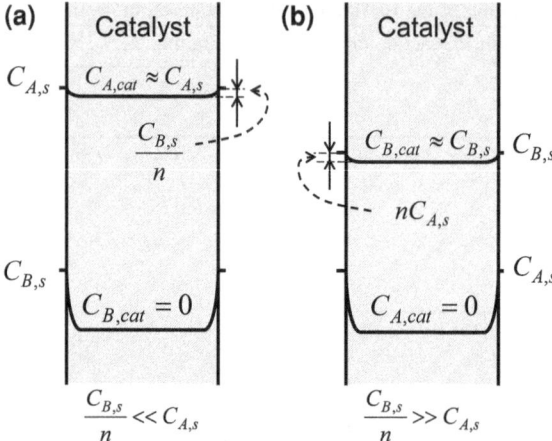

Fig. 3.1 Liquid and gas concentration profiles in the catalyst bulk when a catalyst is active. **a** Extreme case of a gas shortage. The gas concentration falls to zero in a thin region near to the particle shell. The concentration of the liquid reactant also decreases but only by the value equal to $C_{B,s}/n$, which is negligible compared to $C_{A,s}$ so that the concentration of liquid reactant can be considered to be approximately constant throughout the catalyst bulk. **b** Extreme case of a gas excess. Contrary to (**a**), the concentration of the liquid reactant falls to zero resulting in an insignificant drop in the gas concentration by $C_{A,s}n$. In this case, the gas concentration can be regarded constant in the catalyst bulk

The first one is characterized by the low gas concentration $C_{B,s}$ while the concentration of the liquid compound $C_{A,s}$ is high. More exactly, this condition corresponds to $C_{B,s}/n \ll C_{A,s}$ (Fig. 3.1a). The second one is related to the low concentration of the liquid reactant $C_{A,s}$ compared to the gas concentration $C_{B,s}$, which can be expressed as $C_{B,s}/n \gg C_{A,s}$ (Fig. 3.1b) [7].

When the catalyst particle falls under the shortage of the gaseous compound on the catalyst surface, it can cause the catalyst aging. The hydrogen deficiency, as a rule, facilitates the formation of coke, oxidation of the active sites, and so on.

Thus, assigning a task of working up the industrial reactor, which should demonstrate maximally possible productivity or be of a compact design, the following points are conventionally taken into consideration:

(i)   Catalyst should possess high activity and selectivity in a chosen temperature diapason.
(ii)   Catalyst pellets should preferably be as small as possible.
(iii) Operating temperature has to be comparatively high. However, the maximum temperature should always be limited by the appropriate dynamics of the catalyst aging and acceptable selectivity.
(iv)   Concentrations of the reacting compounds on an external catalyst surface should be as high as possible. With regard to the gaseous reactant, it implies its high partial pressure.

3.2 Approaches from the Point of View of External Mass Transfer

In order that the reaction given by Eq. (3.1) can occur, all reacting compounds should be delivered to the catalyst surface by means of external mass transfer. The gas should overcome two mass transfer resistances on the gas–liquid and liquid–solid interfaces while the liquid reactant undergoes only one mass transfer step—liquid–solid (Fig. 3.2).

Generally, there can be two concentration gradients of the gas compound at the gas–liquid interface: gas side and liquid side as shown in Figs. 3.2a and 3.2b. The gas-side concentration gradient can be observed if the gas phase is represented at least by two components including the reacting gas and vapor (or another gas). Such a situation (the existence of the gas-side gradient) can take place when the partial pressure of vapor is significant; for example, when the liquid reactant fed into the reactor is dissolved in a volatile solvent (e.g., hydrogenation of nitro-compounds in the presence of methanol) or the initial liquid reactant and/or its product are volatile (e.g., hydrogenation of acetone). If the partial pressure of the reacting gas is just slightly less than the total pressure, the gas-side gradient can be neglected.

In the general case (Fig. 3.2a), the molar flux of the reacting gas through gas–liquid and liquid–solid interfaces is given by the following equations:

$$j_B = \beta_{B,g}(C_{B,g} - C_{B,g}^*) \tag{3.8}$$

$$j_B = \beta_{B,l}(C_{B,l}^* - C_{B,l}) \tag{3.9}$$

$$j_B = \beta_{B,s}(C_{B,l} - C_{B,s}) \tag{3.10}$$

In terms of Henry's law, the gas–liquid equilibrium at the gas–liquid interface can be written as

$$C_{B,g}^* = HC_{B,l}^* \tag{3.11}$$

From Eqs. (3.8), (3.9), and (3.11), one can yield

$$j_B = K_{g-l}(C_{B,g}/H - C_{B,l}) \tag{3.12}$$

where K_{g-l} is the total gas–liquid mass transfer coefficient normalized to the concentration in the liquid phase

$$K_{g-l} = \left(\frac{1}{H\beta_{B,g}} + \frac{1}{\beta_{B,l}}\right)^{-1} \tag{3.13}$$

Assuming the equality of the specific areas of gas–liquid and liquid–solid interfaces (only for the case of simplicity), one can obtain from Eqs. (3.10) and (3.12)

Fig. 3.2 Concentration profiles of liquid and gas compounds. **a** The real concentration of the gas reactant with a gap at the gas–liquid interface. **b** The gas-side concentration of the gas reactant normalized according to the gas–liquid equilibrium

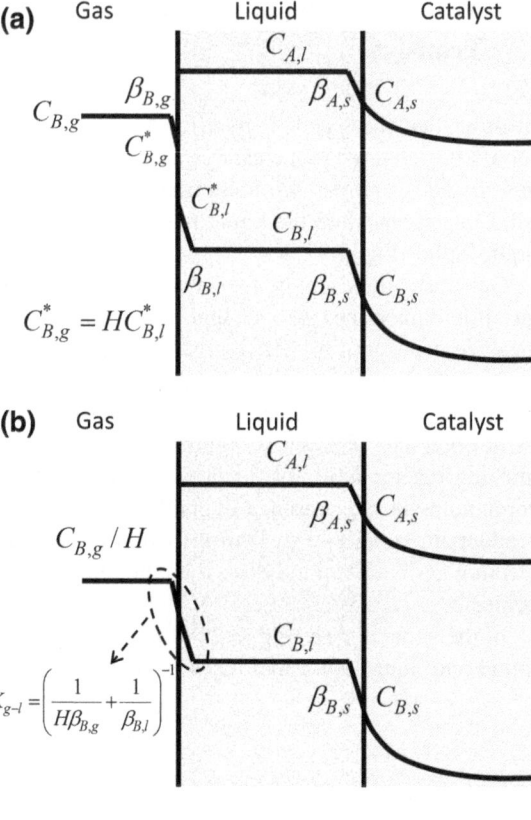

$$j_B = K_{g-s}(C_{B,g}/H - C_{B,s}) \tag{3.14}$$

where K_{g-s} represents the overall gas–liquid–solid mass transfer coefficient

$$K_{g-s} = \left(\frac{1}{K_{g-l}} + \frac{1}{\beta_{B,s}}\right)^{-1} \tag{3.15}$$

The molar flux of the liquid reactant j_A and its relation to the gas molar flux j_B can be expressed as

$$j_A = \beta_{A,s}(C_{A,l} - C_{A,s}) \tag{3.16}$$

and

$$j_A = \frac{j_B}{n} \tag{3.17}$$

Equation (3.18) reflects the dependence of the reaction rate of the liquid reactant r_V upon the molar flux j_A or j_B

$$r_V = a_s j_A = a_s \frac{j_B}{n} \tag{3.18}$$

The bulk concentration of the liquid reactant $C_{A,l}$ varies during the passage of the reactor. In the upstream of the catalyst bed where the fresh liquid enters the reactor, $C_{A,l}$ is maximal and equal to $C_{A,0}$. In the downstream of the catalyst bed, the concentration of the liquid reactant approaches its minimal value $C_{A,out}$, corresponding to the high conversion demanded in the industry. Appropriately, the concentration of the liquid reactant on the external surface of the catalyst particle $C_{A,s}$ varies alongside the catalyst bed from the maximum to minimum.

When the active catalyst is used (i.e., the case of a fast reaction), the concentration of the gas compound on the external catalyst surface $C_{B,s}$ depends on the corresponding concentration of the liquid reactant $C_{A,s}$ [7]. If $C_{A,s} \gg \frac{C_{B,s}}{n}$ (Fig. 3.1a), the surface concentration of the gas compound $C_{B,s}$ becomes nearly equal to zero because of the tremendous surplus of the liquid reactant. That means that the overall reaction rate r_V depends only on mass transfer of the reacting gas according to Eq. (3.19):

$$r_V = \frac{a_s}{n} j_B = \frac{a_s}{n} K_{g-s} \frac{C_{B,g}}{H} \tag{3.19}$$

In the conventional literature (e.g., [8–12]) as well as among process engineers and developers, it is often presupposed that in an overwhelming number of industrial three-phase reactions where active catalysts are employed, catalysts should suffer from a shortage of the reacting gas all over the reactor (i.e., $C_{B,s} \approx 0$). In other words, the reaction rate is always limited by external mass transfer of the gas to the catalyst surface as given by Eq. (3.19). (Note: Below in Sect. 5.5, we will show that this statement is not correct for many applications).

Assuming the case of an active catalyst, the following request on providing appropriate mass transfer with respect to the compact reactor design can be formulated:

(i) Driving force of gas–liquid–solid mass transfer being expressed as $C_{B,g}/H$ must be as high as possible. That implies high pressure in the reactor or a choice of the appropriate solvent, in which the gas solubility is higher (i.e., H is lower).
(ii) Overall gas–liquid–solid mass transfer coefficient K_{g-s} has to be possibly high. That can be achieved by a greater "disturbance" of fluids at gas–liquid and liquid–solid interfaces.
(iii) Liquid–solid mass transfer coefficient $\beta_{A,s}$ has to be also possibly high if mass transfer of a liquid reactant is a limiting stage.

In order to provide the substantial intensity of mass transfer according to cases (ii) and (iii), there must be a sufficient shear velocity between gas and liquid or liquid and solid at the corresponding interfaces, which, in industrial reactors, can be achieved by comparatively high velocities of gas and liquid flowing through the catalyst bed.

For the following consideration, it is necessary to point out that any convective mass transfer is always coupled with the energy consumption into the system: more energy should be delivered if more intensive mass transfer is needed.

3.3 Approaches from the Point of View of Heat Evolution

Even if the two approaches mentioned above are completely fulfilled (i.e., a catalyst of an extreme activity is used and ideal conditions for mass transfer are provided), one essential problem related to the heat production should be solved in industrial reactors. As shown in Table 1.1, many reactions represented by Eq. (3.1) are extremely exothermic. Table 3.2 demonstrates that in some reactions, an adiabatic temperature rise ΔT_{ad} can reach several thousand degrees Celsius (see Sect. 5.1 for more details).

However, as pointed out in Sect. 3.1, the process temperature should be confined to a comparatively narrow range preconditioned, on the one hand, by the appropriate reaction rate and, on the other hand, by the process selectivity and safety. It can only be achieved if the reaction heat is withdrawn from the reaction zone.

It seems that the best and simplest option to take the reaction heat out of the catalyst mass is the integration of a heat exchanger in the catalyst bed by placing the catalyst inside or outside heat transfer tubes (e.g., multitubular reactors or reactors with embedded heat exchangers).

However, such a method cannot alone cope with a huge amount of heat evolved even in reactions of low or medium exothermic effects (e.g., hydrogenation of alkenes).

This statement can be illustrated by uncomplicated evaluations (Appendix A) with respect to the reaction of 1-hexene hydrogenation to n-hexane taken as an example. If this reaction were realized in a fixed-bed reactor of an industrial scale (e.g., the catalyst bulk volume is presumably more than 1 m^3), at least at the initial part of the catalyst bed (where there is a high concentration of 1-hexene), the ratio of the heat transfer surface to the catalyst bulk volume should be enormous.

Such an embedded heat exchanger demands a great volume of the heat transfer tubes, several cubic meters, with their total length of several kilometers. It is obvious that even if such an extremely complex reactor is designed, it will not be exploitable because of high investment and operating costs. It is necessary to point out that there are other crucial points related to the integration of heat exchangers and catalysts in one embodiment. Chapter 5 will elucidate some additional concerns over such a method pertaining to temperature gradients in the catalyst bed, fouling the heat transfer surfaces, flow uniformity, and temperature control.

Thus, the following aspects of the heat transfer in the reactor design should be considered:

Table 3.2 Thermal and energy characteristics of some processes (possible by-reactions are not considered)

N	Process	ΔT_{ad} (K)	ΔT_{permit} (K)	$\dfrac{N_{g,recycle}}{Q_{l,feed}}$ (mol$_g$/m^3_l)	$\dfrac{N_{g,recycle}}{N_{B,feed}}$	$\dfrac{E_{recycle\,f}}{Q_{l,feed}}$ (GJ/m^3_l)
1	Hydrogenation of acetone to isopropyl alcohol	620[a]	70	0.5×10^6	40	0.3
2	Hydrogenation of nitroparaffins (C12-C14) to aminoparaffins	1,100[a]	15	5×10^6	430	3.0
3	Hydrogenation of dinitrotoluene (mixture of isomers) to toluenediamine (mixture of isomers)	4,000[a]	15	17×10^6	390	10.0
4	Hydrogenation of 3,4-dichloronitrobenzene to 3,4-dichloroaniline	900[b]	15	3×10^6	730	1.8
5	Hydrogenation of dinitrotriethylbenzene to diaminotriethylbenzene	440[c]	15	2.0×10^6	350	1.2
6	Hydrogenation of nitrobenzene to aniline	2,500[a]	15	10×10^6	370	5.9
7	Hydrogenation of 2,4-dinitroaniline to triaminobenzene	850[d]	15	3.7×10^6	400	2.2
8	Hydrogenation of furfural to furfuryl alcohol	400[a]	20	1.3×10^6	100	0.8
9	Hydrogenation of furfural to tetrahydrofurfuryl alcohol	1,200[a]	20	4.1×10^6	100	2.4
10	Hydrogenation of 4-nitrosophenol to 4-aminophenol	105[e]	15	0.4×10^6	30	0.24
11	Hydrotreating reactions (incl. hydrocracking)	40–600 (or higher for pyrolytic oils)	10–20	0.04×10^6– 2.9×10^6	2.5–10 [15, 16]	$0.024 \div 1.7$

[a] Without solvents
[b] 30 % solution in toluene
[c] 30 % solution in methanol
[d] 30 % solution in dimethylformamid
[e] 10 % solution in ethanol
[f] Estimated for pure hydrogen at $T_1=363$ K, $P_1=70$ bar, and $\Delta P_\Sigma=20$ bar

(i)   Possibly gradientless temperature profile in all directions.
(ii) Simple and reliable temperature control.
(iii) Possibility of preventing the reactor runaways without any external guard system.

In the framework of the conventional concepts, it will be impossible to achieve these three goals, if some special measure for the heat withdrawal is not undertaken. Such a measure is the use of a reacting gas as a cooling agent. In this case, the temperature rise between inlet and outlet of the reactor is defined by a quantity of the gas going through the catalyst bed: the lower temperature difference corresponds to the greater amount of gas. It is not difficult to ensure the reactor operation in the chosen temperature interval even without any heat exchangers inside.

3.4 Approaches from the Point of View of Hydrodynamics

It is easy to comprehend that all the goals enumerated in Sects. 3.1–3.3 can only be achieved by an appropriate structure of a two-phase gas–liquid flow. Namely, in order to provide the maximum possible reactor productivity (or compact design), the concentrations of reacting compounds on a catalyst surface should be possibly high. That, in turn, necessitates the intensive mass transfer, which can be realized by a comparatively high velocity of the gas flow through the catalyst bed at the relatively high pressure.

At the same time, the effective heat removal can only be provided by an appropriate mass of reacting gas (as a cooling agent) going through the catalyst bed. As will be shown below (see Sect. 5.1), this amount of gas should significantly exceed what is consumed by the reaction itself so that the part of the not converted gas must be taken from the reactor exit and directed again to its inlet.

Thus, the intensive flow of the reacting gas through the catalyst bed has to facilitate the fulfillment of all the goals with regard to kinetics, mass, and heat transfer.

The necessity of injecting the great amount of gas through the reactor imposes some specific features on both the process conditions and the reactor design and the entire configuration of the reactor unit including peripheral equipment (e.g. loop compressors, heat exchangers and recuperators, phase separators, pumps, etc.).

Two main parameters should be mentioned especially: the operating pressure and the pressure drop that should be overcome by the flowing gas. Both these factors define the process efficiency from a point of view of the reaction performance and the energy input demanded for the reactor functioning.

With regard to the effective and compact reactor design, the following points are conventionally taken into account:

(i) Possibly lower pressure drop, which results in the lower energy consumption and the use of unsophisticated loop compressors;

(ii) Possibly high operating pressure that implies the high concentration of the reacting gas. The high pressure also reduces the volume (but not the mass) of the injected gas, which results in the use of the pipework of a smaller diameter and smaller peripheral components as to save the investment;

(iii) Appropriate velocity of gas, which accounts for the efficient flow pattern, especially, with regard to BCRs, where the two-phase gas–liquid flow is directed upward;

(iv) Uniform gas and liquid distribution.

3.5 Concluding Remarks

Naturally, specialists engaged in the design of multiphase reactors should pay attention not only to kinetics, hydrodynamics, mass, and heat transfer, but also to such important issues as, for example, materials, auxiliary equipment, pipework, and so on. (e.g. [13, 14]). On the whole, the complete design should be aimed at the low investment and operating costs that are chiefly determined by operating pressure, energy consumption, and safety precautions (see Sect. 5.8 where the approximate algorithm for the process design is discussed).

As shown above, there is at least one technical method that allows one to consolidate all the features demanded by the process kinetics, mass, and heat transfer, viz., to make the reacting gas flow through the catalyst bed at high pressure. Comparing the chemical consumption of this gas with the amount that should pass the reactor (Sect. 5.1), one is inevitably forced to draw a conclusion about the return of an unreacted part of this gas back into the reactor with the help of a recycle compressor.

The way of thinking given in Sects. 3.1–3.4 apparently reproduces the argumentation of the people belonging to the beginning and middle of the 1900s, who had to accept the challenge of developing fixed-bed multiphase technologies starting from laboratory experiments to industrial applications. Often, they had to make their decisions intuitively under constraints of very short time and lack of necessary physical and chemical data. It is surprising how they contrived to create technologies that are successfully employed by different industries up till now.

Appreciating their efforts, it is necessary to underline that the method, according to which the conditions for kinetics, mass, and heat transfer are created by the simple gas loop under high pressure, does not imply the best "ideology" for carrying out the multiphase processes. So far, the interpretation of physical and chemical processes inherent in fixed-bed reactors is based on the old ideas introduced in this chapter. For example, among the professionals dealing with three-phase reactors the necessity of high pressure in the reactor is mainly explained by the low gas solubility, or the gas–liquid–solid mass transfer limitation, or the inclination of the catalyst to its aging, which, from today's knowledge seems to be not completely and always correct.

In Chap. 5, the conventional fixed-bed reactors will be revised from the point of view of the most recent knowledge on the multiphase catalysis. Special attention will be paid to the interaction between physical and chemical processes occurring in the catalyst bed.

References

1. R.G. Maiti, K.D.P. Nigam, Gas-liquid distributors for trickle-bed reactors: A review. Ind. Eng. Chem. Res. **46**(19), 6164–6182 (2007)
2. E.W. Thiele, Relation between catalytic activity and size of particle. J. Ind. Eng. Chem. **31**, 916–920 (1939)
3. Y.B. Zeldowitch, To the theory of the reaction on a porous or powder catalyst. Acta Physicochimica USSR, **4**, 583–592 (1939)
4. D. Murzin, T. Salmi, *Catalytic Kinetics* (Elsevier, Amsterdam, 2005)
5. R.E. Hayes, S.T. Kolaczkowski, *Introduction to Catalytic Combustion* (Gordon and Breach Science Publishers, Amsterdam, 1997)
6. A. Kadivar, M.T. Sadeghi, M.M. Gharebagh, Estimation of kinetic parameters for hydrogenation reaction using a genetic algorithm. Chem. Eng. Technol. **32**(10), 1588–1594 (2009)
7. L.B. Datsevich, D.A. Muhkortov, Multiphase fixed-bed technologies. Comparative analysis of industrial processes (experience of development and industrial implementation). Appl. Catal. A **261**(2), 143–161 (2004)
8. J.F. Jenck, Gas–liquid–solid reactors for hydrogenation in fine chemical synthesis, in *Heterogeneous Catalysis and Fine Chemicals II*, ed. by M. Guisnet, J. Barrault, C. Bouchoule, D. Duprez, G. Perot, R. Maurel, C. Montassier (Elsevier, Amsterdam, 1991), pp. 1–19
9. M. Herskowitz, Hydrogenation of benzaldehyde to benzyl alcohol in a slurry and fixed-bed reactor, in *Heterogeneous Catalysis and Fine Chemicals II*, ed. by M. Guisnet, J. Barrault, C. Bouchoule, D. Duprez, G. Perot, R. Maurel, C. Montassier (Elsevier, Amsterdam, 1991), pp. 105–112
10. C.N. Satterfield, *Mass Transfer in Heterogeneous Catalysis* (MIT Press, Cambridge, 1970)
11. H. Gierman, Design of laboratory hydrotreating reactors scaling down of trickle-flow reactors. Appl. Catal. **43**, 277–286 (1988)
12. G. Mary, J. Chaouki, F. Luck, Trickle-bed laboratory reactors for kinetic studies. Int. J. Chem. Reactor Eng. **7**, Review R2 (2009)
13. L. Harwell, S. Thakkar, S. Polcar, R.E. Palmer, Study outlines optimum ULSD hydrotreater design. Oil & Gas J. (2003)
14. L. Harwell, S. Thakkar, S. Polcar, R.E. Palmer, Study identifies optimum operating conditions for ULSD hydrotreaters. Oil & Gas J. (2003)
15. J.H. Gary, G.E. Handwerk, M.J. Kaiser, *Petroleum Refining: Technology and Economics* (CRC Press, Boca Raton, 2007)
16. P.R. Robinson, G.E. Dolbear, in *Hydrotreating and hydrocracking: fundamentals*, ed. by C.S. Hsu, P.R. Robinson. Practical Advances in Petroleum Processing, vol 1 (Springer, New York, 2006), 177–218

Chapter 4
Process Flow Diagram and Principal Embodiments of Conventional Industrial Units

According to the previous chapter, a reactor comprising a fixed catalyst can be imagined as the column or ensemble of tubes, through which the recirculation of a reacting gas is arranged. Since the typical gas velocity in the channels formed by the catalyst particles is relatively high, the flow of liquid should be directed concurrently with the gas phase either downwards or upwards. The reactors, where the first flow pattern is realized, are referred to as Trickle-Bed Reactors (TBRs), the up-flow reactors are referred to as Bubble Column (packed) Reactors (BCRs).

Despite the similar arrangements, every industrial process has its own unique configuration, for instance, concerning the heat recuperation or the treatment of the recycled gas. A rather complex network of heat recovery can be encountered in refineries, e.g., in hydrotreating processes where the liquid and gas fluxes for this purpose can be used not only inside the reactor system itself, but also outside it by the integration of these fluxes with the down or upstream stages. Figure 4.1 displaces a simplified flow diagram of a hydrotreater, TBR with a loop of hydrogen-rich gas, employed by one of the refineries. As can be seen, several heat recuperators (8b) are shared with the stripping process downstream (not presented in Fig. 4.1). The technological scheme for hydrotreating can still be more complicated if, for example, a high-pressure scrubber is installed within the gas loop for the removal of hydrogen sulfide formed in the course of hydrodesulfurization reactions.

In spite of the seeming complexity encountered in different industrial units, the schemes of TBRs and BCRs can be simplified for our analysis. Since the process flow and control in TBR and BCR units do not differ very much from each other, the following description will be related to both types of these techniques.

Figure 4.2a, b represents TBRs and BCRs with gas recirculation. The liquid and gas to be processed are fed by pump 7 and compressor 6, respectively, and mixed with the gas recycled by compressor 5. The two-phase gas–liquid mixture passes heat recuperator 8, where it is heated by the hot flux leaving the reactor. The finish

L. B. Datsevich, *Conventional Three-Phase Fixed-Bed Technologies*,
SpringerBriefs in Applied Sciences and Technology,
DOI: 10.1007/978-1-4614-4836-5_4, © The Author(s) 2012

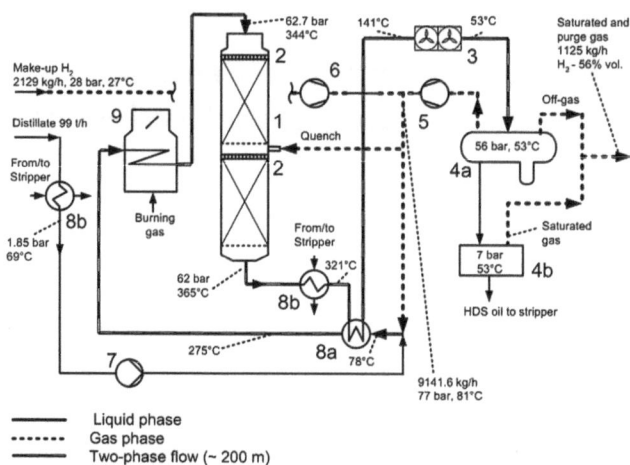

Fig. 4.1 Simplified scheme of a hydrotreater with gas quench. *1* Reactor (∅3.5 m, H = 15 m); *2* Gas–liquid (re)distributors; *3* Cooler; *4a* High pressure phase separator; *4b* Low pressure phase separator; *5* Recycle compressor (ΔP_Σ= 21 bar); *6* Compressor for fresh gas; *7* Feed pump; *8* Recuperative heat exchangers installed in the layout of the hydrotreater (*8a*) and the stripper (*8b*); *9* Fired heater

heating of the gas–liquid flux to the inlet temperature demanded by the process specification is carried out in heater 9.

The reaction between the gas and liquid reactants occurs on a catalyst fixed in reactor 1. Passing the reactor, the gas–liquid flow is consecutively cooled in recuperator 8 and cooler 3. The comparatively cold two-phase mixture enters phase separator 4, from which the liquid product (with dissolved gas) is taken away, while not reacted excessive gas is returned by recycle compressor 5 back to the reactor.

In the course of any reaction, some quantity of by-gases can be formed. In order to exclude the fade of the reaction due to the accumulation of these gases in the gas loop, the purging of the reactor system should be foreseen by setting the appropriate flow rate of the off-gas from phase separator 4.

Temperature and pressure are the main characteristics that should imperatively be controlled.

The temperature regime along the catalyst bed is determined by two parameters. The first parameter is the temperature of the gas–liquid stream at the reactor inlet, and the second is the flow rate of the recycled gas.

The necessary inlet temperature is provided by heat exchanger 9. In some processes, especially in fine chemistry, there can be no heat recuperation. In these cases, only two heat exchangers (3 and 9) are installed. If the temperature level in the reactor should be relatively low (e.g., 30–70 °C as in hydrogenation of some nitroso compounds), heat exchanger 9 should serve as a cooler.

Fig. 4.2 Simplified technological schemes of TBR and BCR (without intermediate quenching). **a** BCR; **b** TBR. *A* liquid reactant; *B* gaseous reactant. *1* Reactor; *2* Gas-liquid (re)distributors; *3* Cooler; *4* Phase separator; *5* Recycle compressor; *6* Compressor for fresh gas; *7* Feed pump; *8* Recuperative heat exchanger; *9* Heater

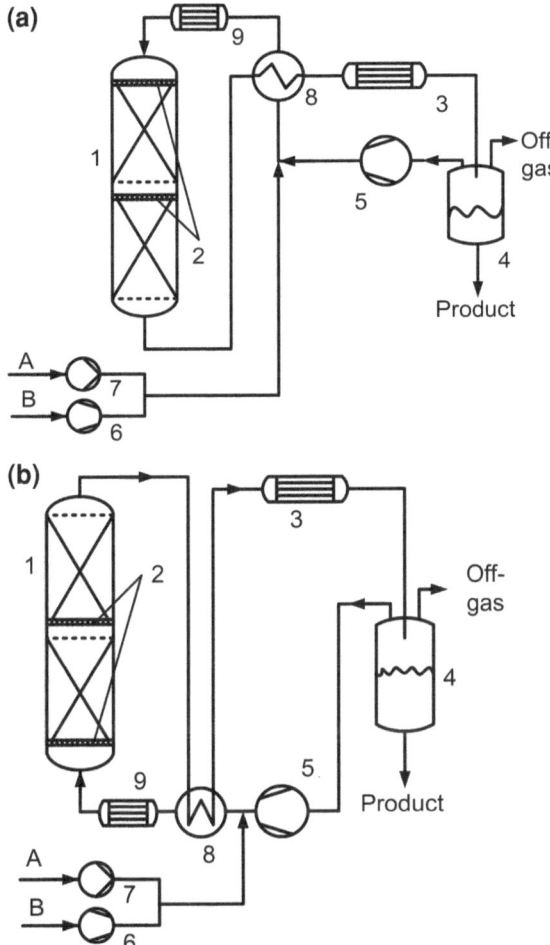

The gas recirculation rate dictates the temperature rise along the catalyst bed. The higher the amount of gas flowing through the reactor, the less the temperature difference should be observed.

Special attention should be paid to the temperature of the recycled gas at the intake and, therefore, at the outlet of recycle compressor 5, which is controlled by cooler 3. As is known, the energy demand for gas compressing depends on the temperature and pressure at the compressor suction (see Appendix B). In order to lessen the energy consumption, the pressure at the intake should be high and the inlet temperature should be as low as possible even if the compressor is designed to withstand higher temperatures. (It is worth pointing out that the discharge temperature, which depends on the suction temperature, can be a subject of legislative norms for some types of compressors [1]). It is also necessary to point out that the lower suction temperature facilitates the higher molar (or mass) flow rate of the recycled gas if all other conditions are equal.

The necessary level of pressure in the reactor system is kept by feed compressor 6, whose task is to hold the constant pressure at its outlet regardless of the gas consumption and purging.

In order to minimize the maldistribution of gas and liquid, usually the fixed catalyst is divided into several beds, each of which is furnished with its own gas–liquid distributor.

Some reactors can be equipped with intermediate gas entrances situated between catalyst beds and designated for the distribution of the recycled gas along the catalyst height as is shown, for example in Fig. 4.1. In refinery processing, the introduction of the part of the cooled recycle gas is called quenching. As a rule, two or three separate beds with one or two quench points are used in hydrotreating units. In fine and bulk chemistry, reactors with more than three catalyst sections with more than two intermediate points for quenching can be encountered.

From the point of view of the process embodiments, the technological schemes of MTRs do not differ from those presented in Fig. 4.2. As is explained in Sect. 3.3, MTRs cannot alone cope with the heat withdrawal to the free space around tubes. That means that the recycle of gas is demanded, albeit with a less flow rate than that compared to TBRs and BCRs.

Since heat transfer in the radial direction is rather bad, the industrial MTRs employ tubes of 4–6 cm in diameter. Even if such small catalyst tubes are used, the temperature difference in the radial direction can be significant, about ten centigrade or more.

It is worthy of special mention that MTRs are very costly because of their complex manufacture. A typical MTR constitutes several dozen thousand tubes, the assembly of which should withstand high temperature and pressure. The maintenance of MTRs is also very complicated especially during catalyst loading: The pressure fall in all tubes has to be proved and equalized in order to prevent bypassing effects at least at the beginning of operation.

As is well known, effective mass and heat transfer in three-phase systems is always characterized by a high degree of the energy input (chiefly mechanical) delivered into the reacting system, for example, by intensive stirring. The "mechanical" energy introduced in the gas–liquid–solid reactors intensifies the interaction between all phases by increasing the phase contacts and the shear stress at their interfaces.

In fixed-bed reactors, the perturbation of the gas and liquid medium inside a catalyst bed is carried out by the gas flow generated by recycle compressor 5.

This compressor is one of the key elements of all schemes because it accounts for the temperature regime, effective mass and heat transfer, as well as for the process economy.

As a rule, single casing, oil-free centrifugal compressors are used for the gas recirculation. The main advantages of these compressors lie in their comparatively small size and untroubled continuous-duty operation. Unfortunately, such compressors have some restrictions, for example, with regard to the pressure head, especially, if gases with low molecular weight like hydrogen are compressed. In some industrial applications, especially in the production of fine chemicals, reciprocating single stage compressors can be encountered.

The typical pressure boost of recycle compressors used in chemical and petroleum industries lies in the comparatively narrow range limited to around 15–20 bar.

The ability of each given compressor to recycle a required amount of gas (in mole or kg) depends on the pressure at the compressor inlet, which should be relatively high. [As will be shown below, in many industrial applications, the choice of the operating pressure is dictated not by the chemical nature of the process, but by the necessity of recycling the demanded amount of gas through the loop system (Sect. 5.2)].

The pressure rise that the recycle compressor should develop has to compensate for the pressure drop in the loop system. This pressure drop represents the hydraulic losses in all elements including pipelines, heat exchangers, the reactor, etc. and can grow during the run because of fouling the loop elements, especially inside the heat exchangers and the catalyst bed.

It is necessary to underline that the pressure drop along each catalyst bed should always be limited because it makes a substantial contribution to the force applied, for example, to the catalyst at the bed support, which should be less than the catalyst crush strength. Usually, the pressure drop over the catalyst bed cannot exceed 2–5 bar.

The energy stored in the compressed gas should provide a necessary "job" in a fixed-bed catalyst, ensuring the intensive mass and heat transfer. To a large degree, however, this energy is dissipated when the recycled gas passes the network of pipelines and heat exchangers either alone or chiefly with the liquid phase. To minimize the energy expenditure in the loop system, the gas should have possibly lower temperature and higher pressure (see Appendix B).

For the further consideration, it is important to emphasize that the whole length of the pipework designated only for the transportation of the gas–liquid mixture can reach several hundred meters. For example, in the hydrotreater presented in Fig. 4.1, this part of the loop pipeline is about 200 m.

Some peculiarities inherent in the transportation of the gas and liquid mixture with regard to the pressure drop and, therefore, to the energy consumption will be detailed later (Sect. 5.7). Here, it is necessary to underline that the overwhelming pressure drop and, as a result, energy demand falls at the loop lines, but not at the reactor. Actually, if 100 % corresponds to the pressure rise produced by the recycle compressor, the loop pipework (including heat exchangers) accounts for the pressure loss by more than 70 %.

Depending on the type of a catalyst, some procedures for its preparation prior to the production start may be needed.

In some processes, the fresh catalyst loaded in the reactor can start running without any special and sophisticated treatment. For example, Pt and Pd or activated Raney catalysts for hydrogenation reactions can operate without any activation although their drying can be demanded.

In hydrotreating units, catalysts are activated in the framework of the same equipment wherein a special mixture of oil and a sulfur-containing compound is fed into the reactor.

However, in many industrial applications, a newly loaded catalyst has to be activated differently. For example, metal oxides in some catalysts (e.g., Ni, Cu, Co, and Fe catalysts) have to be reduced to a metal form by the mixture of hydrogen and nitrogen at 250–320 °C under ambient or higher pressure.

The exothermic gas-phase activation can be carried out with the help of the existing equipment. The heat removal is brought about similar to the usual operation, in heat exchanger 3 by recycling the gas mixture with the help of recycle compressor 5. Activation water is collected after cooling in phase separator 4. If heater 9 is not designed for bringing the gas mixture to high temperatures of 320 °C, an additional heater (often electrical) should also be included in the unit layout.

Other catalysts that need the activation represent alloys between active metals and Al, for instance, Ni–Al catalyst (not activated Raney catalyst). These catalysts should be leached with the water solution of NaOH under 90–110 °C and slightly elevated pressure. After the porous structure is formed, the catalyst should profoundly be washed with water.

As already discussed in Sect. 3.3, if the reaction heat ceases to be removed, the temperature inside the reactor can jump to the level of several thousand degrees. In order to prevent any runaway, some units can be equipped with the special guard system (not shown in Fig. 4.2), the only task of which is to evacuate the reaction mixture from the reactor and fill it with nitrogen.

It is worthy of note that the number of schemes, which can be encountered somewhere, is not encompassed only by those depicted in Figs. 4.1 and 4.2.

For example, the TBRs with the countercurrent flow of gas and liquid or the column reactors with several heat exchangers embedded inside the catalyst bed can be found in the literature. As a rule, such reactors cannot be utilized for strongly exothermic processes.

The countercurrent flow of gas (upwards) and liquid (downwards) cannot be realized because the velocity of gas even under pressure of hundreds bar can be excessively high. Such a significant velocity of gas can lead to the liquid flow inversion and, as a result, to the reactor malfunction, especially if the fouling of the channels between the catalyst particles due to the formation of sticky, high-molecular by-products on the catalyst surface is taken into account.

As for the application of embedded heat exchangers, this point has already been elucidated in Sect. 3.3.

Some additional explanations are likely needed with regard to the use of recycled gas for purposes of intermediate quenching. Although such schemes can be encountered in bulk chemistry, the operating personnel prefer to shut off all intermediate influxes of the recycled gas between the catalyst sections [2]. Hydrotreating and hydrocracking units are the exception to this rule, but some hydrotreaters do not employ quenching all the time, but apply it when there is a temperature increase if, for example, properties of oil feed or pressure drop distribution between sections are changed.

References

1. ANSI/API 618, *Reciprocating Compressors for Petroleum, Chemical, and Gas Industry Services,* 5th edn. (American Petroleum Institute, 2007) washington DC.
2. L.B. Datsevich, D.A. Muhkortov, Multiphase fixed-bed technologies. Comparative analysis of industrial processes (Experience of development and industrial implementation). Appl. Catal. A **261**(2), 143–161 (2004)

Chapter 5
Analysis of Conventional Industrial Processes

When multiphase catalysis in fixed-bed reactors is discussed among the people involved in research and development in chemical and petroleum industries, some misapprehensions stemming from the time of the first fixed-bed reactors always arise. Summing up the numerous discussions, the frequently repeated statements originated from the outdated technological approaches considered in Chap. 3 can be formulated as following myths:

(1) The gas recirculation rate has to be high for providing the greater gas concentration in the liquid phase (for speeding up the reaction rate or slowing down the catalyst aging) or for decreasing the content of by-gases (e.g. hydrogen sulfide in hydrodesulfurization reactors without a high-pressure scrubber installed in the gas loop).

(2) The overall reaction rate in the overwhelming number of industrial processes is limited by mass transfer of the gaseous compound and, therefore, operating pressure has to be high in order to enhance gas–liquid-solid mass transfer.

(3) Operating pressure has to be high in order to increase the concentration of the gaseous compound in the liquid phase and, consequently, the reactor productivity.

(4) The high concentration of the liquid compound at the reactor inlet and high operating pressure have to influence the reactor productivity nearly proportionally to these factors if the reaction rate is of the first order with respect to both gas and liquid reactants.

(5) High pressure in the reactor unconditionally encourages the slower rate of the catalyst deactivation.

In this chapter, we will try to elucidate these points as well as discuss other issues with regard to the process efficiency and safety. Some new ideas related to kinetics and mass transfer in view of the oscillation theory will be presented here as well.

L. B. Datsevich, *Conventional Three-Phase Fixed-Bed Technologies,* SpringerBriefs in Applied Sciences and Technology, DOI: 10.1007/978-1-4614-4836-5_5, © The Author(s) 2012

Fig. 5.1 Temperature control in concurrent TBR and BCR during the catalyst aging. *1* Start-up: $T_0 = T_{min}$; *2* Compensation for the catalyst aging after the start-up: $T_0 > T_{min}$, but $T_0 + \Delta T_{permit} < T_{max}$; *3* End of the catalyst life: $T_0 + \Delta T_{permit} = T_{max}$

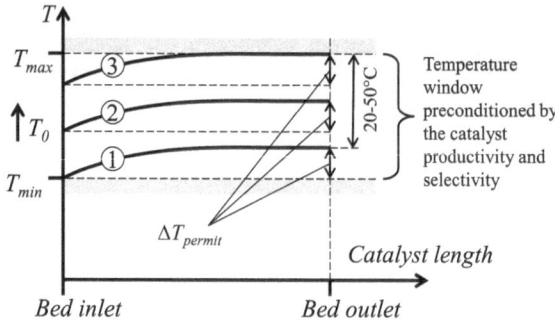

5.1 Temperature Control and Heat Balance

The temperature is the most important parameter that should be strongly controlled. As has been mentioned in Sect. 3.1, the reaction temperature should lie in a comparatively narrow range between T_{min} and T_{max}, where T_{min} implies the start-up temperature that should be high enough to guarantee the appropriate reactor productivity while T_{max} is defined by the demand on the process selectivity and safety.

As a rule, the following strategy for the temperature control is applied to industrial processes (Fig. 5.1).

In the beginning of the operation, when the catalyst is fresh, the temperature of the gas–liquid mixture at the reactor inlet T_0 is set equal to T_{min}.

Because of the exothermic reaction, the temperature of the reacting mixture passing the catalyst bed increases and achieves the maximum value at the reactor outlet in column type reactors with a concurrent flow (TBR and BCR in (Fig. 4.2)). In MTRs, the maximum temperature can be observed somewhere between the reactor inlet and outlet, which depends on the heat transfer properties of an individual reactor tube.

A choice of the admissible temperature rise along the catalyst ΔT_{permit}, which is usually kept constant during the catalyst lifecycle, is based on the necessity to compensate for the catalyst aging during the operation period. As a rule, this temperature rise is preset from a condition to have a substantial temperature reserve: at the beginning of the catalyst cycle $T_0 + \Delta T_{permit}$ should be by 20–50 °C less than T_{max}.

As the catalyst loses its activity, the temperature at the reactor inlet T_0, and, correspondingly, the temperature level in the reactor are increased recouping a loss in the catalyst productivity. This procedure is repeated until the maximum allowable temperature T_{max} is achieved somewhere in the reactor.

Although in MTRs, there is a heat removal through the wall of tubes, the main part of the generated reaction heat is withdrawn by the gas recirculation. In order to avoid the excessive analysis of such reactors with regard to radial heat transfer, we will consider only the column type reactors, keeping in mind the similar features inherent in all reactors with gas recirculation.

Since the heat losses through the wall of the column type reactors (TBR and BCR) can be neglected, the heat balance for these reactors can be expressed as

$$C_{A,0}Q_{l,\text{feed}}X(-\Delta H_A) = \left(\rho_l Q_{l,\text{feed}}C_{P,l} + N_{g,\text{recycle}}C_{P,g}\right)\Delta T_{\text{permit}} \tag{5.1}$$

Here the left-hand side represents heat generation at conversion X whilst the right-hand side reflects the uptake of heat when the gas–liquid flow is heated by ΔT_{permit}. (Note: Eq. (5.1) does not consider partial evaporation of liquid into the gas stream. In many industrial applications, the evaporation process and, therefore, associated heat transfer can be neglected).

From Eq. (5.1), one can obtain the flow rate of recycled gas $N_{g,\text{recycle}}$

$$N_{g,\text{recycle}} = Q_{l,\text{feed}}\left(\frac{C_{A,0}X(-\Delta H_A)}{\Delta T_{\text{permit}}C_{P,g}} - \frac{\rho_l C_{P,l}}{C_{P,g}}\right) \tag{5.2}$$

and the maximum adiabatic temperature rise ΔT_{ad} (defined for complete conversion at the absence of recirculation, i.e. $X = 1$ and $N_{g,\text{recycle}} = 0$)

$$\Delta T_{\text{ad}} = \frac{C_{A,0}(-\Delta H_A)}{\rho_l C_{P,l}} \tag{5.3}$$

From Eqs. (5.2) and (5.3) one can also get the ratio of the recycle gas to the volume flow rate of the liquid feed

$$\frac{N_{g,\text{recycle}}}{Q_{l,\text{feed}}} = \frac{\rho_l C_{P,l}}{C_{P,g}}\left(\frac{\Delta T_{\text{ad}}X}{\Delta T_{\text{permit}}} - 1\right) \tag{5.4}$$

Taking into account that the reactor productivity related to liquid compound A (in mol_A/s) at complete conversion is equal to

$$N_{A,\text{feed}} = Q_{l,\text{feed}}C_{A,0} \tag{5.5}$$

and the relation between the gas and liquid feeds is defined by reaction Eq. (3.1) as

$$N_{A,\text{feed}} = nN_{B,\text{feed}} \tag{5.6}$$

one can obtain the ratio of the gas recirculation rate $N_{g,\text{recycle}}$ to the gas consumed in the course of the reaction $N_{B,\text{feed}}$ as

$$\frac{N_{g,\text{recycle}}}{N_{B,\text{feed}}} = \frac{C_{P,l}}{C_{P,g}}\frac{\rho_l}{nC_{A,0}}\left(\frac{X\Delta T_{\text{ad}}}{\Delta T_{\text{permit}}} - 1\right) \tag{5.7}$$

The parameters evaluated according to Eqs. (5.3, 5.4, 5.7) for some processes are given in Table 3.2.

As can be seen, the exothermic nature of hydrogenation processes demands a huge amount of the recycled gas for the heat withdrawal. This value can be hundred times more than the chemical consumption itself. It is necessary to note that such a great recirculation rate is dictated exclusively by the necessity of the heat removal and has nothing in common with Myth 1.

For the practical applications, the essential question arises: How can such a great mass of gas be forwarded through the loop system?

5.2 Loop Hydraulics as the Main Factor for a Choice of the Operating Pressure

As a rule, the available scientific literature (see, for example, [1, 2]) chiefly focuses on the hydraulic behavior of a catalyst bed. Very seldom, if ever, the role of other elements of the gas loop is analyzed with regard to such parameters as, for example, operating pressure or limitations posed on the catalyst volume in a single reactor.

In this Chapter, we will try to demonstrate that the process pressure in the industrial units is preconditioned rather by the banal hydraulic considerations than by the traditional ideas related to kinetics, mass transfer and hydrodynamics reflected in Chap. 3 and the above-mentioned myths.

The total pressure drop over the whole loop ΔP_Σ is equal to the boost pressure developed by the recycle compressor and includes the pressure losses in the reactor, pipework, heat exchangers, valves, control devices and other apparatuses.

The upper level of the pressure drop that can be allowed for the catalyst bed is predetermined by the maximal force that can mechanically destroy the catalyst. This force is applied to the catalyst in the place of its contact with a supporting grid or gas–liquid distributor.

In a TBR with a downward gas–liquid flow, this force is exerted to the catalyst particles by the grid situated below the catalyst bed.

In order to prevent the particle motion due to the upward flow in BCRs, the catalyst has to be squeezed between the gas–liquid distributor (situated below) and the supporting grid (situated above) so that the site of the maximal force can be localized either beneath or atop the catalyst bed.

The maximal force applied to the catalyst can be defined as an interaction between the pressure difference over the catalyst bed, weight of the catalyst (including liquid holdup as well as the weight of coke and other species that can be formed and deposed in or around the catalyst particles during the operation cycle). In BCRs additionally, the squeezing force should be taken into account.

In all cases, this force should not exceed the catalyst crush strength. Typical pressure drop over the reactor bed with commercial catalysts cannot exceed 2–5 bar and practically lies in the range of 0.7–3 bar.

Taking into account that the commercial gas recycling compressors cannot develop the pressure rise ΔP_Σ more than about 20 bar, let us estimate pressure requirements for the gas loop.

When the demanded flow rate of the liquid feed $Q_{l,\text{feed}}$ is preset, the molar flow rate of the recycled gas $N_{g,\text{recycle}}$ is also preconditioned according to Eqs. (5.2 or 5.4).

Principally, all elements of the gas loop should be analyzed with regard to the pressure loss. In order not to overburden the text, let us evaluate only the pressure drop just over a pipeline section.

According to technological schemes depicted in Figs. 4.1 and 4.2, the pipework encompasses two parts, through one of which only the gas phase flows while the mixture of gas and liquid is forwarded through another. The whole length of the pipework of typical industrial units can reach several hundred meters, the overwhelming part of which is designated for the gas–liquid flow. For example, the pipeline for the two-phase flow in the hydrotreater with a catalyst volume of 114 m^2, the scheme of which is presented in Fig. 4.1, has the length of about 200 m.

The two-phase pressure drop in tubes depends on the character of the flow (upward, downwards or horizontal), gas void fraction, liquid and gas densities and their velocities [3]. In comparison to the pressure drop built by the flow of the single gas phase, the two-phase pressure drop of the same gas flow but together with liquid is considerably greater.

For the following exemplification, we purposefully consider only the gas flow through the loop pipework keeping in mind two following points. First, the dependence of the pressure drop on the gas flow rate and pressure is the same for other loop elements. Secondly, just only the pipeline (tubes) without any other devices like valves, control elements, etc., as a rule, account for 10–20 % of the total pressure fall while the reactor can constitute 3–30 %. Other loop elements like heat exchangers, separator(s), valves and control devices account for a loss in 50–87 %.

As is well known, the pressure drop differential dP built by a single gas phase flowing through a tube section of diameter d_{tube} with a length increment dL_{tube} can be calculated as

$$\frac{dP}{dL_{tube}} = \frac{1}{d_{tube}} \zeta(\text{Re}) \frac{\rho_g U_{tube}^2}{2} \tag{5.8}$$

where $\zeta(\text{Re})$ is the friction factor depending on the Reynolds number $\text{Re} = \frac{\rho_g U_{tube} d_{tube}}{\eta_g}$ and surface roughness, ρ_g and U_{tube} are the density and velocity of gas, respectively.

Inserting into Eq. (5.8) the velocity of gas $U_{tube} = \frac{4N_{g,recycle}RT}{\pi d_{tube}^2 P}$ and gas density $\rho_g = \frac{P\mu_g}{RT}$, one can obtain

$$P\frac{dP}{dL_{tube}} = \zeta(\text{Re}) \frac{8\mu_g RT}{\pi^2} N_{g,recycle}^2 \frac{1}{d_{tube}^5} \tag{5.9}$$

and

$$\text{Re} = \frac{4N_{g,recycle}\mu_g}{\pi d_{tube}\eta_g} \tag{5.10}$$

Table 5.1 Pressure drop (in bar) over a 100 m section of a smooth, uniform pipe in the case of a hydrogen flow at 100 °C

Average pressure P^* (bar) Piping size d_{tube} (cm)	20	50	100	150	200
3	$-$ [a]	$-$ [a]	$-$ [a]	147.1	110.3
5	$-$ [a]	37.0	18.5	12.3	9.2
7	18.1	7.2	3.6	2.4	1.8
9	5.3	2.1	1.1	0.7	0.5
11	2.0	0.8	0.4	0.3	0.2

[a] Flow is not realizable

In the isothermal case, the right-hand side of Eq. (5.9) is independent of pressure and can be integrated to get the pressure drop over a tube section of length L_{tube} as

$$\Delta P = \frac{1}{P^*} \zeta (\mathrm{Re}) \frac{8 \mu_g RT}{\pi^2} N_{g,recycle}^2 \frac{1}{d_{tube}^5} L_{tube} \qquad (5.11)$$

Here P^* corresponds to the arithmetical mean of the inlet and outlet pressures at the ends of the tube section of the length equal to L_{tube}.

Since the friction coefficient $\zeta (\mathrm{Re})$ at the developed turbulence has a slight dependence on the Reynolds number, it is figured out that the pressure drop ΔP is practically proportional to the molar flow rate of recycled gas $N_{g,recycle}$ to the power of two, proportional to the temperature T and inversely proportional to the tube diameter d_{tube} to the power of five. It can be expected that the pressure drop in a two-phase flow has the same (or still more powerful) dependence on the gas flow rate and pipe diameter.

Taking hydrogenation of furfural to tetrahydrofurfuryl alcohol as an example of the reaction with comparatively moderate heat generation, let us estimate the pressure fall in the pipework of 100 m if the production feed $Q_{l,feed}$ is comparatively low and corresponds to 1 m^3 of furfural per h. According to Table 3.2, the flow rate of the recycled hydrogen $N_{g,recycle}$ should be 1.14×10^3 mol/s.

The pressure drop of the single gas flow as a function of pipe sizing and pressure calculated according to Eq. (5.11) are given in Table 5.1.

Let us estimate the operating pressure if the boost pressure ΔP_Σ developed by the recycle compressor is equal to 20 bar and if the industrial unit is equipped with 100 m of pipework, the pressured drop through which makes up 20 % of the total pressure ΔP_Σ, i.e. 4 bar.

From the data in Table 5.1, one can define that the operating pressure in the reactor should be above 100 bar if, for instance, the pipeline with internal diameter of 7 cm is used. If the operating pressure is less than 100 bar, it becomes impossible to cram the demanded amount of the recycled gas through the loop of this piping size. In turn, the heat removal and, consequently, the temperature control in the reactor become unfeasible.

It is necessary to point out that the use of the pipework of a larger diameter can improve the situation only to some extent because it is difficult to change the geometry of other loop elements, for example heat exchangers. Moreover, the limitations posed on the costs of the pipelines, valves and control devices installed on the loop should be taken into account if a new plant is designed. As a rule, their costs grow disproportionately with an increase in their size.

In some cases, the investment in the pipework of a greater diameter (with valves and other devices) can be so high that two tubes are applied instead of one. Sometimes instead of a single reactor, several reactors each with its own loop but of a smaller diameter are designed, as for example, it was made in the production of tetrahydrofurfuryl alcohol with the total feed rate of 1 m^3/h.

Although the estimations given in this Chapter deal only with the gas flow in pipelines, the similar dependence of the pressure drop upon the gas flow rate $N_{g,recycle}$ can be expected for other loop components.

As can be seen from Eq. (5.11), at constant ΔP_Σ the increase in the pressure allows one either slightly to lessen the pipe diameter ($d_{tube} \sim 1/\sqrt[5]{P^*}$) or to raise the molar flow rate of the recycled gas ($N_{g,recycle} \sim \sqrt{P^*}$). The latter also implies the possible enhancement of the reactor productivity if the active catalyst is applied. However, the higher pressure also results in a disproportionate investment.

So far, in this Chapter, none of the specific problems pertaining to kinetics and mass transfer has been discussed with regard to the operating pressure. Nevertheless, proceeding only from three parameters such as (i) reaction heat ($-\Delta H_A$), (ii) allowed temperature rise ΔT_{permit} and (iii) requested volumetric flow rate of feed $Q_{A,feed}$, the minimal operating pressure is exactly preconditioned by the available industrial compressors and the geometry of reactor(s) and loop elements. Moreover, the same considerations directly impose the limit upon the minimum size of catalyst particles, which can be used in fixed-bed reactors.

Thus, the choice of the operating pressure and the productivity of a single reactor lie in the limits dictated by the loop hydraulics and appropriate investment costs and has nothing in common with Myths 2 and 3.

5.3 Macrokinetic Peculiarities

The real chemical process on a single catalyst particle can differ from the classical description given in Sect. 3.1. In the reaction with heat or gas evolution, the oscillatory motions of liquid in catalyst pores with velocities of several hundred meters a second can occur [4–9].

According to the oscillation model, such motions are brought about by the impossibility of removing heat and/or gas by the thermal conductivity or molecular diffusivity, which results in the recurrent appearance of the gas or gas–vapor bubbles in the catalyst pores.

Specifically, a gas–vapor embryo has to come into existence at some point inside a catalyst pore when the total pressure of the saturated gas and vapor at this

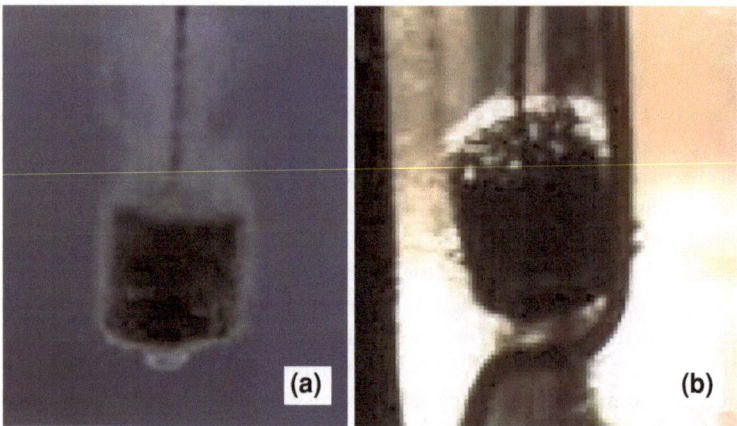

Fig. 5.2 Bubbles in the pore mouths. **a** Hydrogen peroxide decomposition on a 6 × 6 mm Ni catalyst at 25 °C and 30 % of H_2O_2 in H_2O (frame taken from Movie 1 [10]) **b** Hydrogenation of 1-hexene to n-hexane on a 6 × 6 mm Ni catalyst at 216 °C and 80 bar (frame taken from Movie 6 [11])

point becomes equal to the maximally possible pressure in this pore that represents the sum of the capillary pressure and the pressure in the reactor. Due to over-heating and/or oversaturation alongside this newborn bubble, the embryo begins rapidly to grow displacing the liquid out of the pore. On the part of the pore occupied by the bubble, the reaction ceases. Because of the dissipation processes, the temperature and/or the gas partial pressure return to the initial level so that the liquid outside the catalyst pellet begins to fill the pore again driven by the capillary force. The process of the bubble formation, liquid displacement and pore filling repeats as long as the heat or gas is generated. In detail, the mechanism and criteria of the oscillation existence are described in [4, 5, 7].

In many cases, the growing bubble can be so big that it can appear in the pore mouth so that some part of it can be observed from the outside. Figure 5.2a illustrates the bubbles escaping pore mouths in the course of hydrogen peroxide decomposition on a Ni catalyst. This reaction represents the process with the simultaneous heat and gas generation.

It is worth pointing out that if the reaction mixture is intensively disturbed by a stirrer or flow, the vortexes over the pore mouths will tear off the extremely small bubbles coming out of the pores so that their observation becomes nearly impossible [8].

In contrast to the reaction with gas generation, where the oscillatory motion of liquid can be observed by a naked eye in uncomplicated reactors (e.g. Fig. 5.2a), the oscillations in the course of exothermic reactions cannot be so easily detected because of experimental difficulties. Actually, gas–liquid exothermic reactions such as, for example, hydrogenation processes demand sufficient gas–liquid-solid mass transfer, which can be provided by stirring in the experimental set-up. Unfortunately, stirring breaks away small bubbles from the pore mouths so that they cannot be seen even if a transparent reactor cell operating under pressure is used. That is why many scientists doubt that the oscillations predicted by the oscillation theory are really take place in the course of multiphase exothermic reactions. Nevertheless, a specially adjusted laboratory set-up and experimental procedure [9] have recently allowed one distinctively to observe the oscillatory performance even under intensive agitation. In Fig. 5.2b, the appearance of bubbles in the pore mouths of a Ni catalyst is seen in the reaction of hydrogenation of 1-hexene to n-hexane.

Generally, the intensity of oscillations and, therefore, the reaction rate depend on the physical and chemical properties of the reacting system as well as on the pore structure. According to the oscillation theory, the larger the pores are, the more violent motion of liquid can be expected. In fixed-bed reactors, where each particle has several contacts with other pellets, the sites of such contacts can be considered as the pores of extremely big diameters so that the oscillatory motion at these points can be extremely intensive.

Although the oscillations do not change the character of the chemical interaction of reacting species on active catalyst sites, they can drastically affect the reaction behavior on a macro scale.

The following conclusions stemming from the oscillatory behavior are important for the analysis of the industrial processes:

(i) The Thiele/Zeldovich model may not be adequate for the process description because of the different nature inherent in mass and heat transfer in the catalyst particles.

(ii) The effective diffusion coefficients under oscillations are significantly higher than defined by the conventional approaches [4–6]. Moreover, they should depend on the reaction rate. Because of intensive internal convections, the reaction rate should be remarkably higher than that computed by the traditional methods. Furthermore, due to the stochastic character of oscillations, the deep penetration of the reacting compounds in the catalyst bulk should take place, which, in turn, accelerates the reaction rate.

(Principally, the catalyst particle working in the industrial reactor has no necessity of "knowing" whether its behavior is described by the Thiele model or the oscillation theory. It becomes crucial for the scale-up procedure and analysis of the reactor operation, especially, with regard to safety aspects).

(iii) Liquid and gas phases moving in pores alternately with great velocities destroy the laminar sublayer around the catalyst particle resulting in the acceleration of external mass (and heat) transfer. Since the intensity of the oscillation directly depends on the rate of gas or heat generation, external mass (and heat) transfer has to be governed not only by the hydrodynamic conditions of the surrounding gas and liquid flow as it is traditionally accepted, but also by the reaction rate. In fact, the contribution to mass transfer brought by the reaction can significantly outweigh the hydrodynamic impact.

(iv)   The oscillatory motion of liquid can cause cavitation inside pores due to very strong pressure alternations resulting in the mechanical destruction of catalyst particles (see Movie 3 [12]). That, in turn, can necessitate the downstream filtration of the product escaping the reactor.

(v)   In the part of the pore, where the bubble appears, non-evaporating substances can block active sites of a catalyst. That can lead to the initial deactivation of the catalyst during first hours of its lifecycle, for example, as it is observed in hydrogenation of some nitro compounds [4, 5].

(vi)   Considering the space between catalyst particles as small channels and the sites of the particle contacts with each other as pores of large diameters, the appearance of great bubbles outside the catalyst particles can take place, especially, if the rate of heat or gas generation is considerable [4]. The conglomerate of such big bubbles can isolate the part of the catalyst bed from the reacting species so that the reaction ceases. Due to the dissipation of heat (in exothermic reactions) or gas (in reactions with gas evolution), the liquid returns to the catalyst surface and the process of macropulsation begins again.

In the course of the routine operation of industrial reactors, the macropulsations do not probably exist, or, at least, they cannot be detected. However, if the function failure or reactor runaway happens, such macropulsations can be not only registered by temperature sensors, but also distinctly heard.

It is worth underlining that the numerous industrial exothermic processes are accompanied with the co- or by-products that are gaseous or prone to evaporation under operating conditions (e.g. H_2O and NH_3 in hydrogenation of nitro compounds, CO_2 and CO in hydrogenation of furfural, CH_4, C_2H_6, etc. in hydrotreating and hydrocracking). Such gaseous or vaporous products should additionally enhance the oscillatory performance.

Another problem that should be elucidated is the concentrations of the reacting compounds at the part of the catalyst situated near the reactor entrance. The physical and chemical processes on this initial part of the catalyst bed can play an important role with regard to safety. In the overwhelming industrial applications even if the operating pressure is high, the concentration of the liquid reactant entering the reactor is considerably greater than the concentration of the gas compound with regard to the stoichiometric ratio given by Eq. (3.1), i.e.

$$C_{A,l} \gg \left(\frac{C_{B,g}}{H} \right) / n \qquad (5.12)$$

If there is a shortage of the gaseous reactant according to Eq. (5.12), the liquid reactant practically will have the same concentration across the catalyst bulk. At the same time, if the catalyst is active, the concentration of the gas will vary from its maximal value to zero. Moreover, the concentration of the gas compound on the catalyst surface $C_{B,s}$ can be taken equal to zero (Fig. 3.1a). (The exception is the purification reactors that will be considered in Chap. 6).

More precisely, if the gas concentration falls from its maximal level of $\dfrac{C_{B,g}}{H}$ to zero, the concentration drop of the liquid reactant becomes equal to $\left(\dfrac{C_{B,g}}{H}\right)/n$.

For several processes, Table 5.2 shows that the liquid reactant in the catalyst bulk maintains the extremely high concentration at the beginning of the catalyst bed, the value of which is only slightly less than that in the liquid bulk.

5.4 Phenomenology of Gas–Liquid Flow

The detailed description of hydrodynamics of the gas and liquid flow through a catalyst bed can be found in the vast scientific and technical literature (e.g. [13]) so that this chapter purposefully deals only with some peculiarities relevant to the current topic and does not pretend to be all-encompassing narration.

As is said in Chap. 4, in TBRs and BCRs, the liquid and gas are distributed over each catalyst bed by gas–liquid distributors in an attempt to ensure the uniform hydrodynamic conditions everywhere.

As a rule, the superficial velocity of liquid U_l lies in the range of 0.01–1 cm/s while the velocity of gas U_g can reach values of several dozen centimeters a second. (The relation between U_l and U_g can easily be determined proceeding from the ratio $N_{g,recycle}/Q_{l,feed}$ (see Table 3.2) and operating pressure.)

Practically in all fixed-bed reactors, the hydrodynamic pattern has a resemblance to the film flow. In TBR with low gas velocities, the mobile liquid film is formed by the capillary effect and this film is driven downwards by the gravity. If the velocity of gas is comparatively high, the liquid film is additionally forwarded downwards by the shear stress applied by the gas flux to the gas–liquid interface.

In BCR, the climbing film is performed by the gas moving with high velocity in a core of the channel built by the catalyst particles so that the liquid is spread over the catalyst surface. Such a flow pattern resembles the annular flow observed in channels (e.g. see Movie [14]).

The thickness of the liquid film formed around the catalyst particle can be evaluated proceeding from the dynamic liquid holdup and specific external surface area of the catalyst bed and lies in the range of 0.1–0.5 mm. The simplest approach to the estimation of the film thickness can be found somewhere [15].

It is worthy of special mention that despite the thorough efforts to provide the uniform distribution of liquid, the maldistribution cannot be ruled out. Experiments carried out in reactors of the industrial scale show that the non-uniformity always exists and increases in the direction of flow [16, 17] due to channeling and segregation [18].

Furthermore, the properties of the catalyst bed in many industrial processes change during the operation. For example, in hydrogenation of some nitro compounds, furfural and other substances, one can observe the irregularities in the

Table 5.2 Mass transfer characteristics of some industrial processes

N	Reaction	Process conditions and production scale	$\frac{C_{B,g}}{H}$ (mol/ m³)	$C_{A,0}$ (mol/m³)	Liquid surplus $\frac{C_{A,0}}{\frac{1}{n}\left(\frac{C_{B,g}}{H}\right)}$	Concentration fall of the liquid reactant across the catalyst at the reactor inlet $\Delta C_A = \frac{1}{n}\left(\frac{C_{B,g}}{H}\right)$	Concentration drop of the liquid reactant in percentage $\frac{\Delta C_A}{C_{A,0}} \times 100\%$	Gas surplus $\frac{\frac{1}{n}\left(\frac{C_{B,g}}{H}\right)}{C_{A,0ml}}$	$\frac{\beta_{B,g}}{\beta_{A,s}}$	$C_{A,inv}$ (mol/ m³)	$\frac{I_{inv}}{I_{bed}}$
1	Hydrogenation of nitroparaffins (C12–C14) to aminoparaffins $CH_3-(CH_2)_i-NO_2+3H_2 \rightarrow$ $CH_3-(CH_2)_i-NH_2+2H_2O$	150 bar 90 °C X = 0.99 Industrial	780	3490	14	260	7	7	3.7	960	0.4
2	Hydrogenation of 3,4 dichloronitrobenzene to 3,4 dichloroaniline (dichloronitrobenzene $+ 3H_2 \rightarrow$ dichloroaniline $+ 2H_2O$)	200 bar 100 °C X = 0.998 Industrial	710	1570 (30 % mass in toluene)	7	240	15	75	2.5	680	0.2
3	Hydrogenation of dinitrotriethylbenzene to diaminotriethylbenzene (dinitrotriethylbenzene $+ 6H_2 \rightarrow$ diaminotriethylbenzene $+ 4H_2O$)	100 bar 100 °C X = 0.99 Pilot-plant	540	940 mol/m³ (30 % mass in methanol)	11	90	10	10	3.3	294	0.4

catalyst bed when the industrial reactor is opened to discharge the catalyst after its lifecycle. These irregularities can be distinguished by the color and density of the tar-like deposition formed around catalyst particles. Sometimes one can get the impression that the catalyst particles are glued with each other so that space between them is blocked by the clammy mass. It is a reason to suppose that in the catalyst bed, there can be places where some conglomeration of catalyst particles can partly be bypassed by the gas flow so that the heat generated inside this zone cannot be sufficiently removed.

Apparently, the same can take place in MTRs. Due to uneven deposition of high-molecular compounds, some tubes can suffer from the lower velocity of the gas flow. Moreover, if such high-molecular substances are formed during the reaction, they are apt to be condensed on the cooled surface of tubes, which additionally worsens their heat transfer properties.

Let us underline once more that in the parts of the catalyst bed situated not far away from the liquid inlet, the concentration of the liquid reactant in the liquid and catalyst bulk is considerably high. That can cause some troubles attributed to the process safety. Namely, what can happen if this initial part of the catalyst bed with a high content of the liquid reactant gradually becomes to be isolated from the gas flow by depositions? The answer on this question is given in the analysis of runaways below (Sect. 5.8).

The change in the bed properties stemming from the depositions can also lead to an increase in the pressure drop across the reactor. If the flow rate of the liquid feed remains the same and the loop compressor operates at its full capacity, the amount of the recycled gas will be decreased. That inevitably increases the temperature in the reactor beyond the admissible level, which, in turn, accelerates the formation of deposition and catalyst aging.

5.5 Mass Transfer

5.5.1 Mass Transfer Coefficients

The comprehension of industrial processes and scaling-up procedures demand the knowledge of mass transfer coefficients (Sect. 3.2).

Usually, gas–liquid mass transfer coefficients are determined in adsorption or desorption experiments. Liquid–solid mass transfer coefficients are obtained in the course of the dissolution of solid particles having the shape of an industrial catalyst and made, for instance, from naphthalene, β-naphthol, benzoic acid, etc. During mass transfer experiments, the hydrodynamic pattern of a gas and liquid flow is adjusted similar to that in industrial reactors. The dependence of the mass transfer coefficients upon the hydrodynamic parameters and physical properties of the tested compounds are then summarized in the form of mass transfer correlations.

There are a great number of papers, where different gas–liquid and liquid–solid mass transfer correlations are presented. Some of the reviews devoted to such correlations can be found somewhere [19, 20]. In spite of the discrepancies in the correlations produced by different authors, there are some common tendencies observed by all research groups regardless of liquids and gases used in their trials.

For gas–liquid mass transfer, such correlations are usually expressed as

$$Sh_{B,l} = \frac{K_{g-l}d_{\text{cat}}}{D_{B,l}} = f(d_{\text{cat}}, D_r, \varepsilon)\text{Re}_g^{n_1}\text{Re}_l^{n_2}Sc_{B,l}^{0.5} \tag{5.13}$$

where function $f(d_{\text{cat}}, D_r, \varepsilon)$ introduces the influence of the reactor and catalyst geometry.

Although the mass transfer coefficients based on these correlations diverge very much in their values, some similarities inherent in them can be pointed out. All correlations for both TBR and BCR indicate a very weak dependence of gas–liquid mass transfer coefficients on a gas velocity. Actually, power index n_1 lies in the range of 0.09–0.4. The liquid flow has a more essential impact, n_2 is found to be in the range of 0.32–0.7.

Since the mechanism of mass transfer of liquid and gas compounds from the liquid bulk to the catalyst surface has the same nature, the identical mass transfer correlations can be used for the definition of mass transfer coefficients $\beta_{A,s}$ and $\beta_{B,s}$.

For TBR, such correlations are mainly presented by equations of the following type

$$Sh_{i,s} = \frac{\beta_{i,s}d_{\text{cat}}}{D_{i,l}} = f(\varepsilon)\text{Re}_l^{n_3}Sc_{i,l}^{1/3} \tag{5.14}$$

where i represent either liquid ($i = A$) or gas ($i = B$) compounds; $f(\varepsilon)$ is a function of the bed geometry.

The index of power n_3 does not vary as much as the analogous value of n_2 in Eq. (5.13) and lies in the range of 0.5–0.822. It is worth pointing out that according to the experiments, the gas velocity does not affect liquid–solid mass transfer.

As to BCRs where the co-current gas and liquid upflow takes place, the expressions for liquid–solid mass transfer can be more complex than Eq. (5.14). However, the liquid–solid mass transfer coefficients also have a weak dependence on the gas velocity.

The faint impact of the gas velocity on liquid–solid mass transfer coefficients under conditions typical in industrial TBRs and BCRs can be explained by the fact that the disturbances of the gas–liquid interface generated by the gas flow decay in the liquid laminar sublayer built over the catalyst surface.

It is noteworthy that the traditional studies on external mass transfer are based on the assumption that the reaction has no effect on the mass transfer mechanism, which is conventionally formulated in terms of the categorical independence of mass transfer correlations from the reaction rate. This is precisely the reason of why the mass transfer correlations produced in the "inert" dissolution or absorption experiments are always applied to the description of the real reactors without any doubt.

In light of the new knowledge about the oscillatory mechanism, this approach seems to be not correct. The exothermic reactions as well as reactions with gas generation may have a great impact on external mass transfer. Actually, the liquid chaotically moving in the catalyst pores with velocities of 100 m/s or more and gas–vapor bubbles escaping the pore mouths should break the liquid film around the catalyst particles resulting in the considerable intensification of mass transfer. Since the intensity of the oscillatory motion is dependent upon the reaction rate, the reaction rate should directly affect the external mass transfer.

The influence of the chemical reaction on external mass transfer was stated in the past only in two papers [21, 22], but failed to be interpreted properly.

The first guess that the reaction could cause vortices of unknown nature in the liquid around the catalyst particles was referred to hydrogenation of α-methyl-styrene on a Pd/Al$_2$O$_3$ catalyst [21].

In the process of acetone hydrogenation in a TBR [22], the dependence of external mass transfer on the reaction rate was investigated in order to design an industrial reactor. It was shown that the gas–liquid-solid mass transfer coefficients were by an order of magnitude greater than values based on the conventional correlations. Moreover, it turned out that the mass transfer coefficients did not depend on the liquid velocity. In that work, this phenomenon was explained by the Marangoni effect, viz.: by eddies generated by the surface tension gradient. The Sherwood number was expressed by the sum of two terms. The first one represented the conventional correlations like Eqs. (5.13) or (5.14) whereas the second one reflected the contribution of the chemical reaction.

The same approach can be applied to the oscillatory mechanism. Provided that a reaction impact can prevail over the influence of hydrodynamic parameters, external mass transfer will become insensitive to a variation of gas and liquid velocities.

Let us emphasize once more that the conventional correlations for TBRs and BCRs demonstrate a very faint impact of the gas flow on gas–liquid mass transfer. Furthermore, the gas flow does not practically affect liquid–solid mass transfer. In the framework of the oscillatory performance, the gas flow should have still less effect.

5.5.2 Two-Zone Model for an Active Catalyst

As has been discussed in Sect. 3.2, the conventional approach to reactor design assumes that the industrial processes are nearly always performed under the limitation of gas mass transfer. Actually, the maximum possible (equilibrium) concentration of a reacting gas in the liquid is far lower than the concentration of a liquid reactant in the liquid bulk, especially, if the stoichiometric relation given by Eq. (3.1) is taken into consideration.

Prima facie, the data presented in Table 5.2 confirm this conclusion. As can be seen, at the reactor inlet there is a considerable excess of the liquid reactant over the gas compound with regard to the reaction stoichiometry. This surplus of the liquid reactant denominated as *L.S.* can be expressed according to Eq. (3.1) as

$$L.S. = C_{A,0} / \left[\frac{1}{n} \left(\frac{C_{B,g}}{H} \right) \right] \tag{5.15}$$

As is seen in Table 5.2, the liquid reactant substantially exceeds the gas compound. It inevitably leads to a pseudo evident conclusion reflected in Myth 2 and Sect. 3.2 that the reaction in TBRs and BCRs with a gas loop should always be limited by external mass transfer of the gas compound if the catalyst is active.

In order to puzzle out this myth, let us analyze the concentration profiles of the reacting compounds on the external catalyst surface alongside the bed of the active catalyst.

The situations illustrated in Table 5.2 are common for many industrial applications, especially in bulk and fine chemistry. Namely, there is a great surplus of the liquid reactant in the part of the catalyst bed nearby the feed entrance. That implies that the concentration of the gas compound on the catalyst surface should be equal to zero if the catalyst is active.

However, the inverse case can be observed at the reactor outlet.

The typical industrial processes are characterized by the high conversion of a liquid reactant, which permits a producer either to arrange the cost-effective separation of the product or to avoid the separation stage at all. This means that at least somewhere near the reactor outlet, the concentration of the liquid reactant in the liquid bulk $C_{A,l}$ can be so low that the excess of the gas compound over the liquid reactant can be found.

The excess of the gas compound over the liquid reactant in the liquid bulk at the reactor outlet can be defined analogically to Eq. (5.15) as

$$G.S. = \frac{1}{n} \left(\frac{C_{B,g}}{H} \right) / C_{A,\text{out}} \tag{5.16}$$

Table 5.2 indicates that at conversions requested in industrial processes, the amount of the reacting gas can have a significant surplus over the liquid reactant. Therefore, if an active catalyst is used, the gas excess results in the negligible concentration of the liquid reactant on the catalyst surface, i.e. $C_{A,s} \approx 0$.

Thus, the bed with an active catalyst can be divided into two zones. The first one corresponds to a gas shortage on the catalyst surface

$$\begin{aligned} C_{B,s} &\approx 0 \\ C_{A,s} &> 0 \end{aligned} \tag{5.17}$$

and the second one represents a shortage of the liquid reactant

Fig. 5.3 Concentration
profiles of gas and liquid
compounds in the liquid bulk
along the catalyst bed

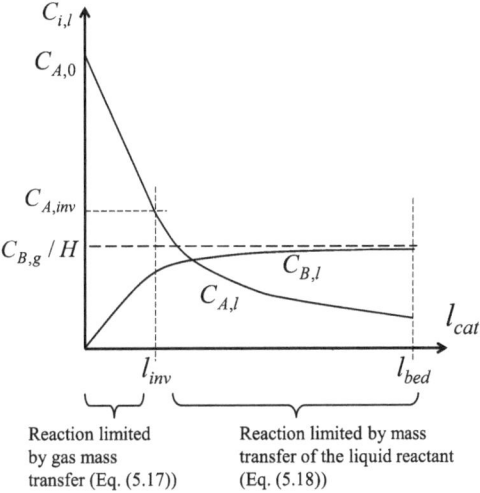

Reaction limited   Reaction limited by mass
by gas mass   transfer of the liquid reactant
transfer (Eq. (5.17))   (Eq. (5.18))

$$
\begin{aligned}
C_{A,s} &\approx 0 \\
C_{B,s} &> 0
\end{aligned}
\tag{5.18}
$$

In the first part of the catalyst bed, the overall reaction rate is limited by mass transfer of the gas compound, whereas in the second part, the total reaction rate is determined by mass transfer of the liquid reactant (see Fig. 5.3).

The inversion point l_{inv} between these two regions should correspond to the situation when both reacting compounds have low concentrations on the external catalyst surface, viz.: $C_{A,s} \approx 0$ and $C_{B,s} \approx 0$.

The concentration of the liquid reactant in the liquid bulk $C_{A,inv}$ at this inversion point can be determined proceeding from Eqs. (3.14), (3.16) and (3.17) as

$$
C_{A,inv} = \frac{1}{n} \frac{C_{B,g}}{H} \frac{K_{g-s}}{\beta_{A,s}}
\tag{5.19}
$$

or taking into account Eq. (3.15)

$$
C_{A,inv} = \frac{1}{n} \frac{C_{B,g}}{H} \left(\frac{\beta_{B,s}}{\beta_{A,s}} \times \frac{1}{1 + \frac{\beta_{B,s}}{K_{g-l}}} \right)
\tag{5.20}
$$

Equations (5.19) and (5.20) are received under the simplification that the interface areas between gas–liquid and liquid–solid phases are equal. It is pertinent to note that this oversimplification cannot bring a great discrepancy in our analysis.

The concentration of the liquid reactant $C_{A,inv}$ can be estimated taking into account that, on the one hand, under conditions typical in industrial reactors, as a rule, $\beta_{B,s}/K_{g-l} < 1$ and, on the other hand, according to the dependence of liquid–solid mass transfer coefficients on the molecular diffusivity according to Eq. (5.14),

$$\frac{\beta_{B,s}}{\beta_{A,s}} = \left(\frac{D_{B,l}}{D_{A,l}}\right)^{2/3} \tag{5.21}$$

If the value of $C_{A,\mathrm{inv}}$ is calculated according to Eqs. (5.20) and (5.21), the length of the catalyst bed from the reactor inlet to the inversion point l_{inv} can be estimated and compared with the entire length of the catalyst bed l_{bed}.

According to Eq. (3.14), in the part of the catalyst bed $0 < l_{\mathrm{cat}} < l_{\mathrm{inv}}$ (see Fig. 5.3), it is valid

$$-U_l \frac{dC_{A,l}}{dl_{\mathrm{cat}}} = \frac{1}{n} a_s K_{g-s} \frac{C_{B,g}}{H} \tag{5.22}$$

or

$$l_{\mathrm{inv}} = (C_{A,0} - C_{A,\mathrm{inv}}) \times \left(\frac{a_s K_{g-s}}{U_l} \frac{1}{n} \frac{C_{B,g}}{H}\right)^{-1} \tag{5.23}$$

or taking into account Eq. (5.19)

$$l_{\mathrm{inv}} = \frac{U_l}{a_s \beta_{A,s}} \left(\frac{C_{A,0}}{C_{A,\mathrm{inv}}} - 1\right) \tag{5.24}$$

In the part of the catalyst bed $l_{\mathrm{inv}} < l_{\mathrm{cat}} < l_{\mathrm{bed}}$ (see Fig. 5.3) according to Eq. (3.16)

$$-U_l \frac{dC_{A,l}}{dl_{\mathrm{cat}}} = a_s \beta_{A,s} C_{A,l} \tag{5.25}$$

Integration of it produces

$$l_{\mathrm{bed}} - l_{\mathrm{inv}} = \left(\frac{U_l}{a_s \beta_{A,s}}\right) \times \ln\left(\frac{C_{A,\mathrm{inv}}}{C_{A,\mathrm{out}}}\right) \tag{5.26}$$

Combining Eqs. (5.24) and (5.26), one can obtain

$$l_{\mathrm{bed}} = \left(\frac{U_l}{a_s \beta_{A,s}}\right) \times \left[\left(\frac{C_{A,0}}{C_{A,\mathrm{inv}}} - 1\right) + \ln\left(\frac{C_{A,\mathrm{inv}}}{C_{A,\mathrm{out}}}\right)\right] \tag{5.27}$$

or

$$\frac{l_{\mathrm{inv}}}{l_{\mathrm{bed}}} = \left(1 + \frac{C_{A,\mathrm{inv}}}{(C_{A,0} - C_{A,\mathrm{inv}})} \times \ln\left(\frac{C_{A,\mathrm{inv}}}{C_{A,\mathrm{out}}}\right)\right)^{-1} \tag{5.28}$$

Depending on the chemical and physical nature, the initial part of the catalyst bed, where the reaction rate is limited by mass transfer of the gas compound, can be less than 50 % of the entire catalyst length. As is shown in Table 5.2, an overwhelming part of the catalyst bed suffers from the mass transfer limitation imposed by the liquid reactant (not by the gas compound!).

Moreover, such industrial processes as purification or ultra-deep hydrodesulfurization (if hydrocracking reactions can be neglected) cannot have this initial part of the catalyst bed at all (see Chap. 6). In these reactors, the overall reaction rate is exclusively predetermined by liquid–solid mass transfer if the active catalyst is used.

Thus, the idea postulating the key role of mass transfer of the gas compound and reflected in Myth 2 is not correct for many industrial applications.

For the sake of clarity, it is necessary to point out that there are some processes, in which mass transfer limitation of gas (hydrogen) takes place everywhere in the catalyst bed, for example, hydrocracking or severe hydrotreating, or hydrogenation reactions where a comparatively low grade of conversion is requested.

5.6 Specific Productivity of the Catalyst Bed

One of the important characteristics of industrial processes accounting for both investment and operation costs is specific productivity of the catalyst bed. The specific productivity denominated here as $S.P.$ (another equivalent term used mainly in refinery industry is liquid hourly space velocity (LHSV)) is defined as a ratio of the feed flow rate $Q_{l,feed}$ to the bed volume V_{bed} that is demanded for conversion of the liquid reactant A from its initial concentration $C_{A,0}$ to the finishing concentration $C_{A,out}$ requested by the process specifications:

$$S.P. = \frac{Q_{l,feed}}{V_{bed}} \qquad (5.29)$$

Taking into account that $Q_{l,feed} = U_l F_{reactor}$ and $V_{bed} = l_{bed} F_{reactor}$, Eq. (5.29) can be transformed to

$$S.P. = \frac{U_l}{l_{bed}} \qquad (5.30)$$

If the active catalyst bed is applied, the specific productivity can easily be estimated proceeding from the two-zone mass transfer model considered in Sect. 5.5.

Inserting the value of l_{bed} from Eq. (5.27) into Eq. (5.30), one can obtain

$$S.P. = a_s \beta_{A,s} \times \left[\left(\frac{C_{A,0}}{C_{A,inv}} - 1 \right) + \ln\left(\frac{C_{A,inv}}{C_{A,out}} \right) \right]^{-1} \qquad (5.31)$$

where $C_{A,inv}$ is calculated according to Eq. (5.20).

Let analyze how the operating parameters can influence the specific productivity if an active catalyst is used.

Fig. 5.4 Effect of pressure
on the specific productivity of
the catalyst bed in
hydrogenation of 3,4-
dichloronitrobenzene

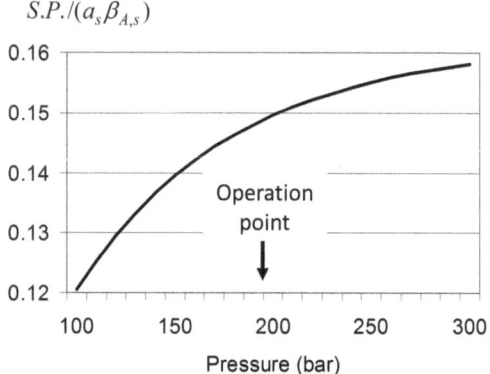

$S.P./(a_s\beta_{A,s})$

As can be seen from Eq. (5.20), the concentration $C_{A,\text{inv}}$ is practically independent of gas and liquid velocities U_g and U_l as well as the particle size d_{cat}.
Actually, $\beta_{B,s}/\beta_{A,s}$ is only defined by the molecular diffusion coefficients of
reactants A and B in the liquid phase (Eq. (5.21)). Even if the value of $\beta_{B,s}/K_{g-l}$ is
not significantly less than unity and, therefore, it should be taken into account, this
ratio has a very slight dependence on U_g, U_l and d_{cat}. This fact follows from the
mass transfer correlations given, for instance, by Eqs. (5.13) and (5.14), viz.:
$\beta_{B,s}/K_{g-l} \sim U_l^{n_3-n_2} U_g^{-n_1} d_{\text{cat}}^{n_3-n_1-n_2}$ where $n_1 \in (0.09; 0.4)$, $n_2 \in (0.32; 0.7)$ and $n_3 \in$
$(0.5; 0.822)$ (see Sect. 5.5.1).

Returning to Eq. (5.31), one can conclude that the specific productivity has no
notable dependence on gas velocity U_g.

The influence of the particle size d_{cat} and the velocity of the liquid phase U_l is
chiefly determined by factor $a_s\beta_{A,s}$. Taking into account that $a_s \sim d_{\text{cat}}^{-1}$ and
$\beta_{A,s} \sim d_{\text{cat}}^{n_3-1} U_l^{n_3}$ in the case of TBRs (Eq. (5.14)), one can yield that

$$S.P. \sim U_l^{n_3} d_{\text{cat}}^{n_3-2} \qquad\qquad (5.32)$$

(It is necessary to point out that if the developed oscillatory behavior takes
place in the course of a reaction, the liquid–solid mass transfer coefficient and,
therefore, the specific productivity have to be independent from the liquid velocity
at least in the part of the catalyst bed where the reaction rate is significant).

The concentration of the gas reactant in the gas phase $C_{B,g}$ and, correspondingly, operating pressure influences the specific productivity through $C_{A,\text{inv}}$ in the
second multiplier in Eq. (5.31). If the vapor pressure is negligible, the concen-
tration of the liquid reactant $C_{A,\text{inv}}$ at the inversion coordinate l_{inv} is proportionally
increased with the operating pressure. Physically, that means that coordinate l_{inv}
that divides the catalyst bed into two zones moves towards the entrance of the
liquid feed. Figure 5.4 shows the influence of the total pressure on the specific
productivity of the catalyst bed in hydrogenation of 3,4-dichloronitrobenzene
calculated according to Eq. (5.31). As can be seen, the total pressure has a

comparatively faint effect on the catalyst productivity if the operating pressure corresponds to the appropriate hydraulic conditions for gas recycling.

The two-zone model allows us also to make a conclusion about incorrectness of Myths 4 and 5. Actually, neither the operating pressure nor the initial concentration of liquid reactant $C_{A,0}$ has the proportional impact on the total reaction rate. Since a great part of the catalyst bed in the reactions of bulk and fine chemistry does not suffer from the shortage of the gas compound, it is impossible directly to connect the catalyst deactivation with low gas concentrations on the catalyst surface.

5.7 Is the Energy Delivered to the Reactor System Dissipated Effectively?

As is well known, the intensive mass transfer in multiphase reactors demands the significant energy consumption for establishing the effective interaction between gas, liquid and solid phases. Energy to the reacting space can be delivered by different methods, for example in slurry reactors, by mechanical stirrers or by a gas or liquid current produced by pumps and compressors. However, the "useful" energy, which is utilized by mass transfer, represents only a part of the total energy spent by the driver (stirrer, pump, compressor, etc.). Much energy is always lost at its transformation, for instance, at friction either in seal packing (slurry reactors with mechanical stirrers) or in pipework if mixing is provided by the current of gas or liquid.

In the fixed-bed reactors with the gas loop, the energy demanded for arranging the appropriate flow pattern, for sufficient mass transfer and heat removal is delivered by the recycle compressor forwarding the gas through the catalyst bed. Table 3.2 indicates the energy introduced by the gas recirculation $E_{recycle}$ related to the flow rate of a liquid reactant $Q_{l,feed}$. It is interesting to explore which part of the energy generated by the recycle compressor is spent "usefully" on the work inside the catalyst bed.

Since in an overwhelming number of industrial reactions, the chemical gas consumption $N_{B,feed}$ is negligible (see Table 3.2), the molar flow rate of the gas recycled through all elements of the gas loop including the reactor can be taken constant and equal to $N_{g,recycle}$. According to Appendix B, the energy dissipated by the gas flow during its passage through the chosen section of the gas loop is practically proportional to the pressure loss over this section while the total energy input in the system is proportional to the pressure boost developed by the recycle compressor.

Thus, if the pressure is monitored in different points of the technological scheme, the pressure drop over the selected element of the gas loop allows one to estimate the energy dissipation in this element.

Fig. 5.5 Energy distribution diagram. **a** The entire reactor system; **b** Catalyst bed *1* Reactor; *2* Recuperative heat exchanger; *3* Cooler; *4* Phase separator; *5* Recycle compressor; *6* Compressor for fresh gas

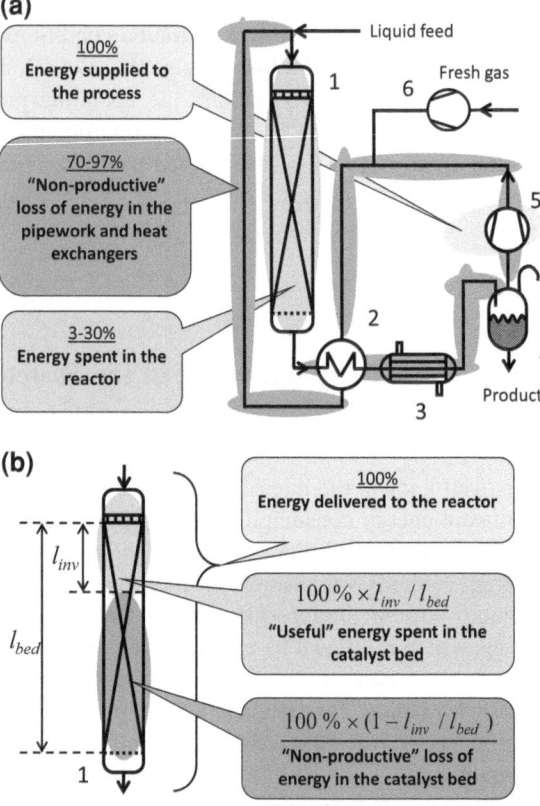

As is pointed out in Chap. 4 and Sect. 5.2, the typical values of the pressure rise produced by the recycle compressors ΔP_Σ lie in the comparatively narrow diapason between 15 and 20 bar. Depending on the reaction, used catalyst, configuration of the technological scheme, geometry of the reactor and gas–liquid distributors, the pressure loss across the reactor $\Delta P_{reactor}$ can vary in the diapason of 0.7–5 bar.

Let the total energy input into the system, i.e. the energy of the gas flow generated by the recycle compressor, be taken as 100 %. Since the energy supply or dissipation is proportional to the corresponding pressure differences, the part of the energy delivered to the reactor or spent in other loop elements can be estimated as the ratio of the corresponding pressure drop to the pressure boost of the recycle compressor, viz.: $\Delta P / \Delta P_\Sigma$.

Figure 5.5a shows that the immense part of the total energy (70–97 %) is mainly spent on the "non-productive" purposes, specifically, on the transportation of the gas and gas–liquid mixture through the loop pipework. Only a small part of the whole energy input equal to 3–30 % falls to the reactor's share providing the "useful" work, namely, mass and heat transfer in the catalyst bed.

Thus, we can conclude that the most industrial processes are characterized by extremely low efficiency with regard to the energy consumed by the reactor.

Moreover, even this "useful" portion of the energy, which is finally delivered to the reactor, is spent not efficiently in the catalyst bed. In order to confirm this point, let us return to Fig. 5.3.

On the initial part of the catalyst bed enclosed between coordinates corresponding to the reactor inlet $l_{cat} = 0$ and the inversion point $l_{cat} = l_{inv}$, the reaction rate is limited by mass transfer of the gas compound. As is mentioned in Sect. 5.5, the gas flow through this part of the catalyst bed affects gas–liquid mass transfer and, therefore, facilitates the reaction performance.

However, in the second part of the catalyst bed situated between the inversion point l_{inv} and the reactor outlet l_{bed}, the reaction rate is limited by mass transfer of the liquid reactant to the catalyst surface, on which the gas flow has no influence (e.g. see Eq. (5.14)). If the gas flow is hypothetically cut off after the inversion point, the reaction rate on this part of the catalyst bed will remain the same. In other words, that implies that the gas flow provides no useful effect in this part of the catalyst bed and the energy accumulated by gas is parasitically dissipated.

The fraction of the energy wasted in the reactor can be easily estimated. If the flow rate of the gas and liquid phases is not changed along the catalyst bed due to evaporation or condensation processes, the pressure will linearly decrease with the length coordinate l_{cat}. That implies that the energy consumed by the first part of the catalyst bed, which can be regarded as "useful" because it provides gas–liquid mass transfer, can be defined as a product of multiplying the energy delivered to the reactor (equal to 3–30 % of the total energy input) by l_{inv}/l_{bed} (see Fig. 5.5b)

Taking into account that l_{inv}/l_{bed} is typically less than 0.4, the energy supply for the reaction performance makes up values less than 1.2–12 % of the total energy input provided by the recycle compressor.

Thus, we can see that from a thermodynamic point of view, the efficiency of the conventional multiphase processes with gas recycling is extremely bad. The overwhelming portion of the energy introduced into the technological scheme by the recycle compressor is not spent for the reaction purposes, but parasitically dissipated in the pipework and the second part of the catalyst bed.

This conclusion is of great importance because it allows us to understand why the conventional fixed-bed reactors have no potential for the process improvement (Chap. 7).

5.8 Process Safety

Although runaways occurring in industrial multiphase reactors are not unique and encountered in many operating companies, there is little information about them in the available literature. Generally, if a runaway happens without injuries or a considerable damage to environment, the analysis of the case will be not disclosed by an operator and will be kept inside its files. Few exceptions (e.g. [23, 24]) convincingly underline this rule.

As is discussed in Sect. 3.3 and 5.1, the most industrial reactions are characterized by extremely high adiabatic temperature rise ΔT_{ad} from several hundred up to several thousand degrees Celsius (Table 3.2), which can principally be achieved in industrial reactors if reaction heat is not removed from the reaction zone.

It is necessary to point out once more that the adiabatic temperatures calculated according to Eq. (5.3) and presented in Table 3.2 are attributed to the goal reactions. The real temperature that can be achieved if the process runs away can be still higher. If the temperature in some part of the catalyst bed oversteps some threshold level and if the gas compound continues to be delivered to the reaction zone, the further reactions between the gas, goal product and solvent will occur and these reactions will run with by far greater heat evolution. In many hydrogenation reactions, such a threshold can be considerably low, about 200–400 °C. It can easily be exceeded if, for example, the molar flow rate of the recycled gas $N_{g,recycle}$ decreases, or the flow rate of the liquid feed $Q_{l,feed}$ as well as the concentration of the liquid reactant in the liquid feed $C_{A,0}$ are increased due to process malfunction.

The influence of $N_{g,recycle}$, $Q_{l,feed}$ and $C_{A,0}$ on the temperature rise ΔT in the column type reactors can be estimated proceeding from the heat balance presented by Eq. (5.1).

If $\Delta T/\Delta T_{ad} \ll 1$, the following dependence can be yielded

$$\Delta T \approx \frac{Q_{l,feed} C_{A,0}}{N_{g,recycle}} \times \frac{(-\Delta H_A)}{C_{P,g}} \tag{5.33}$$

Equation (5.33) indicates that the temperature rise is proportional to the liquid feed and the inlet concentration of the liquid reactant and inversely proportional to the molar flow rate of the recycled gas.

The operating pressure also has an impact on the temperature rise. If the recycle compressor has no capacity margin, i.e. its ultimate power is rigidly adjusted to the process parameters, the decrement of the suction pressure results in the approximately proportional decrease in the gas molar flow rate in the loop.

As is seen in Table 3.2, the permitted temperature growth in the reactors is limited by 10–70 °C that prevents any thermal runaway if the above-mentioned process parameters are kept according to the routine process specifications.

The dangerous temperature excursion can take place if the gas recirculation is shut down or becomes extremely low (e.g. because of a compressor defect or a pressure fall) or the liquid reactant concentration in the feed is increased. Even if the gas and liquid feeds into the reactor are immediately cut off by the control and watching system, the amount of the reacting gas in the whole unit can be enough for developing the hazardous situations. Sometimes it can lead to the burn-off of the reactor wall. An extremely sharp rise in pressure caused by evaporation can result in the reactor destruction if the pressure safety valve fails or does not adequately respond to the rapid evaporating dynamics.

It is important that the resumption of the flow rate of the recycled gas or initial pressure just after a short interruption may not cope with the temperature excursion. If the temperature in the reactor runs beyond the threshold, the gas recirculation will contribute the supply of the reacting gas to the by-reactions with by far higher heat evolution. This heat production can be so high that it can by no means be withdrawn with the help of the recycled gas if its flow rate and, therefore, the recycle compressor are adjusted to a demand on the goal reaction.

At the malfunction of the gas recirculation, the ignition of the catalyst may begin in the part of the catalyst bed situated near the liquid entrance because the concentration of the liquid reactant in this region is extremely high (see Sect. 5.3). The amount of the liquid reactant and the reacting gas (held up in the reactor and in pipework) can be sufficient to trigger the sharp temperature jump. It is necessary to point out that the rapid temperature growth at a runaway can happen during an extremely short time, which can be explained by the oscillation mechanism. Namely, when the liquid is still present in the catalyst pores, heat and vapor evolution accounts for the fast dynamic characteristics due to the extremely high internal (effective) diffusivity (Sect. 5.3). The following scenario is realized: The hotter the catalyst gets, the faster it becomes hot.

It is necessary to note that such a self-accelerating mechanism can be additionally enhanced if some decomposition reactions are taken into account, for instance, the decomposition of nitro compounds. The energy release of such reactions can be about 2000 kJ/mol [25], which is about four times more than the heat evolution of the goal reaction in the production of corresponding aromatic amines.

It seems that if all operating parameters are steady, no runaway can occur. Unfortunately, the emergency case can arise even if there is no observable change in the process parameters.

As is mentioned in Sect. 5.4, the clammy and gummy-like compounds formed around the catalyst particles can be detected in industrial and pilot-plant reactors. In fact, these formations fill the channels built by the catalyst pellets. The gas flow through these channels becomes impeded and less than requested.

If, for instance such a situation happens in one of the numerous tubes of a MTR, the flow rate of the gas through this tube will get less and fail to remove the generated heat, which inevitably leads to the temperature increase in this specific tube.

The similar situation can take place in the column type TBRs and BCRs. Some part of the catalyst particles glued by polymer-like substances begins suffering from the low velocity of the recycled gas. If in this part there is a high concentration of the liquid reactant, the insufficient gas flow will provoke the catalyst heating beyond the permitted level, initiating the temperature excursion.

Such a scenario is likely to account for one of runaways happened in hydrogenation of furfural. In the production of tetrahydrofurfuryl alcohol, one of the reactors operating under 150 bar ran away. The temperature sensors placed inside the catalyst bed not far away from the reactor inlet showed an extremely sharp temperature rise. Although all blockings were activated at 170 °C, the maximum temperature reached during this excursion was higher than 600 °C. Unfortunately, the temperature could not be monitored thereafter because 600 °C was an upper range limit of the recording system.

The fast temperature jump was accompanied with the instant pressure growth that could not be released by the safety-pressure assembly. Despite the fact that the reactor was authorized to the pressure of 300 bar, the top reactor-vessel head was torn off and thrown away in a distance of several hundred meters.

The analysis of all process diagrams recorded prior to the runaway showed that there were no deviations in the process parameters. The similar progress in the temperature excursion is analyzed in benzene hydrogenation [24].

Therefore, it is not surprising that many industrial units are furnished with a special guard system, the only purpose of which is to evacuate the reaction mixture by purging the reactor with a neutral gas, nitrogen.

5.9 Approximate Algorithm of the Research and Design Procedures

As a rule, the kinetic data (e.g. in the form of Eq. (3.2)) can be obtained in the reactor, where a great stirring intensity ensures the absence of concentration gradients in the liquid bulk so that the concentrations of reactants on the catalyst surface can easily be defined in virtue of their equality to those in the liquid bulk.

Another important source of the macrokinetic information is the pilot-plant or laboratory unit, which is in common a small replica of a future industrial process. Although the hydrodynamic conditions (velocities of gas and liquid, pressure drop, gas and liquid distribution, wall effects, etc.) cannot, as a rule, be reproduced in the pilot-plant unit because of much less scale than that of an industrial reactor, the data received in the pilot-plant or laboratory tests are always of great importance.

(i)   These data indicate the approximate span of the process temperature.
(ii)   They allow a researcher or developmental engineer to estimate the specific productivity of the catalyst and the process selectivity in the chosen range of the process parameters.
(iii) The experiments in the pilot-plant or laboratory unit can enlighten the researcher about the rate of catalyst deactivation.
(iv)   The strategy for the compensation for a loss in the catalyst activity by a temperature rise can be evaluated.
(v)   The representative samples of the product can be collected and handed over for quality tests.

The main project decisions are based on the following design procedures.

First, the amount of gas (kg or mole) to be recycled is defined proceeding from the allowable temperature rise, which has to be not too high (chief criteria for the temperature increase are such rival, competing features as productivity, selectivity and rate of catalyst aging). For column type TBRs and BCRs, this temperature growth does not depend on a catalyst, pressure, etc. and is only a characteristic of the reaction heat and desired productivity (naturally, if assumed that conversion and selectivity are close to 100 %). For MTRs, the external heat transfer should result in the lower molar flow rate of the recycled gas.

The next step encompasses the estimation of the catalyst volume necessary for an industrial production output. For this evaluation, the data should be obtained from experimental and pilot-plant runs.

The third step comprises the choice of the reactor geometry, which is mainly associated with manufacturing costs. It is preferably to have possibly lower ratio of the reactor diameter to its height, but, naturally, the catalyst bed height as well as gas velocity should be reasonable.

The fourth step involves the estimation of the pressure drop at gas recycling. This pressure drop depends on the catalyst size, gas recycle rate, geometry of all loop elements and pressure. For example, the higher pressures result in the less pressure drop over the entire loop system so that the more compact and effective recycle compressor is needed.

The pressure drop that any industrial recycle compressor can overcome is limited to about 15–20 bar, which is dictated by the compressor design as well as by economic efficiency of gas compressing and whole processing.

On this stage, the size of a catalyst should be specified taking into account that the smaller the catalyst particles are used, the higher the pressure drop should be overcome by the recycle compressor.

Despite the fact that the small catalyst particles are more preferable for the reaction, their size cannot be less than 1–5 mm because of the pressure drop problem. Moreover, in many processes, the possibility of decreasing the void space between catalyst particles due to the formation of sticky, high-molecular by-products on the external catalyst surface also can lead to an increase in the pressure drop, which should be foreseen in the process layout.

The final step is an attempt to evaluate the efficiency of gas–liquid and liquid–solid mass transfer together with a chemical reaction by methods of mathematical modeling.

Thus, in the course of all research and projecting procedures, the following parameters are defined:

(i) A minimum amount of gas in kg (or mole) to be recycled through the reactor, which cannot be decreased because of process selectivity and safety;
(ii) A size of catalyst particles;
(iii) Reactor geometry;
(iv) A maximum pressure drop, which is impossible to increase because of the economic efficiency, mechanical stability of the catalyst and the absence of recycle compressors available for industrial needs;
(v) Pressure in the reactor, the increase of which results both in positive and negative sequences. Namely, the higher pressure lessens the pressure drop and, to some extent, can improve the catalyst productivity, but it extremely increases the costs of the reactor (thicker wall, special welding procedures and

thorough manufacturing, etc.) and other equipment (valves, tubes, heat exchangers, control devices, etc.). In addition, it should be taken into account that the hydrodynamic pattern (especially in bubble column reactors) can be changed because of lower gas velocity.

In the course of numerous sequential computations aimed at the low investment and operating costs, the design of the reactor and periphery equipment can be accomplished.

As a rule, the gas recirculation rate is nearly always taken to its minimum value that, nevertheless, provides the permitted temperature rise.

As will been shown in Chap. 6, the purification processes without feasible heat generation do not need gas loop. The high gas recycle rate encountered in many industrial applications is the result of numerous misapprehensions shared by process developers.

5.10 Concluding Remarks

So far, many specialists engaged in multiphase catalysis share the conventional design paradigms described in Chap. 3 as well as the myths stemming from them. We hope that our analysis completely disproves these misconceptions.

Actually:

(i)   The high flow rate of the gas in the loop is necessitated by the removal of reaction heat, and not by mass transfer, kinetics or other reasons.
(ii) The high operating pressure has nothing in common with a desire for the greater gas concentration on the catalyst surface and, therefore, for the higher reaction rate. It is dictated by the banal hydraulic reason.
(iii) In many industrial processes, the total reaction rate is mainly limited by mass transfer of the liquid reactant, not by the gas compound. The overwhelming part of the catalyst bed does not suffer from the shortage of the gas compound on the catalyst surface (there are only a few exceptions, for example, hydrocracking reactions).
(iv) The high operating pressure in hydrogenation reactions cannot also be regarded as a decisive parameter accounting for catalyst deactivation (see also Chap. 6) because there is a comparatively high hydrogen concentration on the catalyst surface in the overwhelming part of the catalyst bed (however, higher hydrogen pressure in hydrocracking processes encourages longer catalyst life).
(v)   According to the gas and liquid concentration profiles (see Fig. 5.3), neither the operating pressure (see also Chap. 7) nor the concentration of the liquid reactant in feed has a proportional effect on the overall reaction rate (reactor productivity).
(vi) The reaction rate in the final part of the catalyst bed (Fig. 5.3) cannot be enhanced by increasing the gas recirculation rate because the gas velocity has no impact on liquid–solid mass transfer.

The comparatively new technologies (e.g. GIPKh and POLF processes realized on pilot-plant and industrial scale (Chap. 9)) also confirm these conclusions. They demonstrate the same or significantly higher reactor productivity under the pressure, which is several times less than that in the conventional TBRs and BCRs. At the same time, the deactivation behavior and process selectivity are the same or better.

Our analysis also indicates that the efficiency of the conventional processes with a gas loop is extremely low with regard to the energy dissipation. Even if new catalysts of a greater activity were applied, no essential payoff could be expected (see Chap. 7).

References

1. B. Aydin, F. Larachi, Trickle bed hydrodynamics and flow regime transition at elevated temperature for a Newtonian and a non-Newtonian liquid. Chem. Eng. Sci. **60**, 6687–6701 (2005)
2. R.A. Holub, M.P. Duduković, P.A. Ramachandran, A phenomenological model for pressure drop, liquid holdup, and flow regime transition in gas-liquid trickle flow. Chem. Eng. Sci. **47**(9–11), 2343–2348 (1992)
3. B.A. Shannak, Frictional pressure drop of gas liquid two-phase flow in pipes. Nucl. Eng. Des. **238**, 3277–3284 (2008)
4. L. Datsevich, Oscillations in pores of a catalyst particle in exothermic liquid (liquid/gas) reactions. analysis of heat processes and their influence on chemical conversion, mass and heat transfer. Appl. Catal. A. Gen. **250**, 125–141 (2003)
5. L. Datsevich, Alternating motion of liquid in the catalyst pores at a liquid/liquid-gas reaction with the heat and gas production. Catal. Today **79–80**, 341–348 (2003)
6. B. Blümich, L.B. Datsevich, A. Jess, T. Oehmichen, X. Ren, S. Stapf, Chaos in catalyst pores: can we use it for process development? Chem. Eng. J. **134**, 35–44 (2007)
7. L. Datsevich, Some theoretical aspects of catalyst behaviour in a catalyst particle at liquid (liquid-gas) reactions with gas production: oscillation motion in the catalyst pores. Appl. Catal. A. Gen. **247**(1), 101–111 (2003)
8. L.B. Datsevich, Oscillation theory: Part 4 some dynamic peculiarities of motion in catalyst pores. Appl. Catal. A. Gen. **294**, 22–33 (2005)
9. T. Oehmichen, L. Datsevich, A. Jess, Influence of bubble evolution on the effective kinetics of heterogeneously catalyzed gas/liquid reactions. Part 2: Exothermic gas/liquid reactions. Chem. Eng. Technol. **6**, 921–931 (2010)
10. Movie 1—Oscillatory behaviour in the reaction of hydrogen peroxide decomposition, MPCP GmbH, Illustrative material to the oscillation theory (http://mpcp.de/en/research_and_development/oscillation_model/illustrative_material/)
11. Movie 6—Oscillatory behaviour in exothermic reactions, MPCP GmbH, Illustrative material to the oscillation theory (http://mpcp.de/en/research_and_development/oscillation_model/illustrative_material/)
12. Movie 3—Destruction of a catalyst particle in the reaction of hydrogen peroxide decomposition, MPCP GmbH, Illustrative material to the oscillation theory (http://mpcp.de/en/research_and_development/oscillation_model/illustrative_material/)
13. M.H. Al-Dahhan, F. Larachi, M.P. Dudukovic, A. Laurent, High-pressure trickle-bed reactors: a review. Ind. Eng. Chem. Res. **36**, 3292–3314 (1997)
14. G.F. Hewitt, Gas-Liquid Flow, in *Thermopedia*, Begell House eResourse, (2010) http://www.thermopedia.com/video/toc/images/figs_chaptg/annular doi:10.1615/AtoZ.g.gas-liquid_flow

15. C.N. Satterfield, *Mass Transfer in Heterogeneous Catalysis* (MIT Press, Cambridge, 1970)
16. A.V. Sapre, D.H. Anderson, F.J. Krambeck, Heater probe technique to measure flow maldistribution in large scale trickle bed reactors. Chem. Eng. Sci. **45**, 2263–2268 (1990)
17. V.M. Ramm, *Absorption of Gases (in Russian)* (Khimia, Moscow, 1976)
18. M.H. Al-Dahhan, Y. Wu, M.P. Dudukovic, Reproducible technique for packing laboratory scale trickle-bed reactors with a mixture of catalyst and fines. I&EC Res. **34**, 741–747 (1995)
19. M. Herskowitz, Trickle-bed reactors: a review. AIChE J. **29**(1), 1–18 (1983)
20. P.A. Ramachandran, R.V. Chaudhari, *Three-Phase Catalytic Reactors* (Gordon and Breach Science Publishers, New York, NY, 1983)
21. F. Turek, R. Lange, Mass transfer in trickle-bed reactors at low Reynolds number. Chem. Eng. Sci. **36**, 569–579 (1981)
22. M.I. Nagrodskii, P.N. Ovchinnikov, L.B. Datsevich, Mass Transfer in a Three-Phase Hydrogenation Reactor, in *Intensification of Mass and Heat Transfer in Chemical Apparatuses (in Russian)*, ed. by Y.V. Sharikov, E.I. Leskhina (GIPKh, Leningrad, 1985), pp. 9–15
23. E. Goossens, R. Donker, F. van den Brink, Reactor Runaway in Pyrolysis Gasoline, in *Hydrotreatment and Hydrocracking of Oil Fractions: Proceedings of the 1st International Symposium/6th European Workshop, Oostende, Belgium, February*, ed. by G.F. Froment, B.G. Delmon, P. Grande (Elsevier, 1997), pp. 255–264
24. G. Eigenberger, U. Wegerle, Chapter 12: Runaway in an Industrial Hydrogenation Reactor, in *Chemical Reaction Engineering—Boston, ACS Symposium Series*, vol. 196 (ACS, Washington, 1982), pp. 133–143
25. F. Stoessel, Experimental study of thermal hazards during the hydrogenation of aromatic nitro compounds. J. Loss Prev. Process. Ind. **6**(2), 79–85 (1993)

Chapter 6
Purification Processes

In industrial applications, there are many three-phase processes dealing with the removal of the contaminant presenting in the liquid phase, the concentration of which is insignificant, e.g., hydrogenation of unsaturated compounds or ultra-deep hydrodesulfurization.

Because of a low concentration of the compounds to be treated, such processes have no problem with reaction heat. The reactions practically run under isothermal conditions. Nevertheless, the gas loop with a considerable flow rate of the recycled gas is used.

Hydrogenation of alkenes in the mixture of C5-C6 paraffins can be considered as an example of such applications. (This process represents a recuperation stage in the production of some organosilicon compounds.) The reaction is carried out on an Ni catalyst (4×4 mm) under pressure of 30 bar in a BCR. The concentration of unsaturated compounds is of 0.5 % mass, which corresponds to $C_{A,0} = 50$ mol/m^3 of pentene taken as an equivalent to compounds with double bonds. The equilibrium concentration of hydrogen in the liquid phase at 30 bar is equal to 210 mol/m^3.

The gas–liquid ratio $\frac{N_{g,recycle}}{Q_{l,feed}}$ set by the project specifications is equal to 9.0×10^4 mol$_{H2}$/m^3_l. It has to be pointed out that such a great value is dictated neither by heat evolution nor by mass transfer. Actually, the adiabatic temperature rise calculated according to Eq. (5.3) is just about 4 °C.

Mass transfer also causes no troubles. The reaction mixture entering the reactor becomes saturated with hydrogen in the part of the pipework where gas and liquid move together, i.e., $C_{B,l} = 210$ mol/m^3 at $l_{cat} = 0$. According to Eq. (5.16), there is a 4-fold surplus of the gas compound over the liquid reactant at the inlet of the catalyst bed.

This means that the reaction rate is not limited by mass transfer of the gaseous compound. Consequently, there is no reason for the intensive gas flow through the catalyst bed.

The concentration profiles of the liquid and gas reactants versus the catalyst length are presented in Fig. 6.1. As can be seen, the gas compound has a significant excess in the liquid phase. It results in the extremely high concentration of hydrogen on the external catalyst surface, which can be estimated from Eqs. (3.10), (3.16), and (5.21) as

L. B. Datsevich, *Conventional Three-Phase Fixed-Bed Technologies*,
SpringerBriefs in Applied Sciences and Technology,
DOI: 10.1007/978-1-4614-4836-5_6, © The Author(s) 2012

Fig. 6.1 Typical
concentration profiles of gas
and liquid compounds along
the catalyst bed in
purification reactions

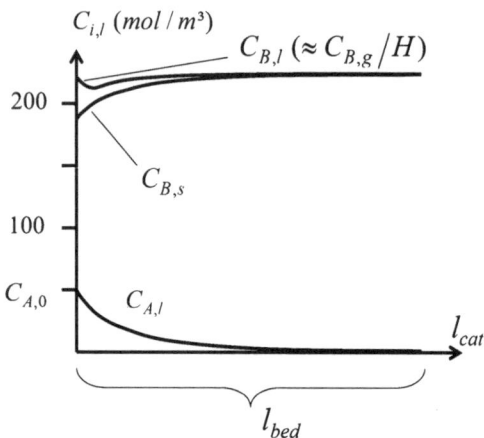

$$C_{B,s} = C_{B,l} - \left(\frac{D_{A,l}}{D_{B,l}}\right)^{2/3} C_{A,l} \tag{6.1}$$

if an active catalyst is used (i.e., $C_{A,s} = 0$), or

$$C_{B,s} = C_{B,l} - 0.43\, C_{A,l} \tag{6.2}$$

The profile of $C_{B,s}$ is also presented in Fig. 6.1. Comparing Figs. 5.3 and 6.1, we see that $C_{A,0} < C_{A,\text{inv}}$ (see Sect. 5.5).

Thus, in purification processes, the overall reaction rate is limited by mass transfer of the liquid reactant, which cannot be affected by the gas flow. Liquid–solid mass transfer can be enhanced if the liquid velocity is increased. However, in the framework of the technological scheme with a gas loop (Figs. 4.1 and 4.2), it is absolutely impossible because of the pressure drop problem.

According to the considerations given in Sect. 5.7, it can be concluded that the efficiency of the energy consumption is extremely poor.

In the author's opinion, the application of the technological schemes with gas recirculation to the purification processes is a design mistake and underlines once more misapprehensions coming from the last century. It is worth mentioning that the analysis given in this chapter has already helped to create other approaches to such processes that were successfully realized [1–3] (Chap. 9).

References

1. L.B. Datsevich, F. Grosch, R. Köster, J. Latz, J. Pasel, R. Peters, T. Pohle, H. Schiml, W. Wache, R. Wolfrum, Deep desulfurization of petroleum streams: novel technologies and approaches to construction of new plants and upgrading existing facilities. Chem. Eng. J. **154**, 302–306 (2009)
2. A. Jess, L. Datsevich, N. Gudde, Purification process, W.O. Patent 0,309,136,3, 2003
3. L.B. Datsevich, M.P. Kambur, D.A. Mukhortov, Hydrotreating in processes of recuperation of spent oils from motors and electric transformers, Fuels and Lubricants (in Russian) 2001, 3 (on line publication http://www.apris.ru/?page=static&section=51)

Chapter 7
Do the Conventional Fixed-Bed Reactors Possess any Potential for the Process Intensification?

If a multiphase catalytic process in a fixed-bed reactor was regarded as a competition between catalyst developers and chemical engineers, nearly always the prize should be awarded to the catalyst developers. Catalysts created by chemists, as a rule, can demonstrate a great ability to do their "chemical" work, if only the process arrangements do not pose impervious limitations toward the desired catalyst performance.

Such limitations can be associated with the insufficient and costly gas recirculation for mass and heat removal, which cannot be overcome because of insurmountable conditions imposed by the loop hydraulics. Although the catalysts loaded in many existing reactors are of greater capability, any attempt to increase their productivity inevitably fails because of both the pressure drop problems in all loop elements and the absence of efficient recycle compressors of higher pressure boost.

Actually, as shown in Sect. 5.6, if the catalyst used in the process is active, there are three parameters that can affect the catalyst productivity: specific velocity of liquid U_l, catalyst particle size d_{cat} and maximum possible (equilibrium) concentration of the gas compound in the liquid phase $C_{B,g}/H$ determined by the operating pressure.

Unfortunately, it is impossible to enhance the reactor productivity by increasing the liquid velocity or by lessening the catalyst size. As explained in Sect. 5.1, any increase in the catalyst productivity should always be compensated by a proportional rise in the molar flow rate of the recycled gas $N_{g,recycle}$ in order to keep the temperature rise in the catalyst bed at the level appropriate for the safe and selective operation.

For example, at the same operating pressure, the doubled productivity of an industrial reactor demands the doubled molar flow rate of the gas, which, in turn, results in at least a 4-fold increase in the pressure drop over the entire gas loop (see Eq. (5.11)). Taking into account that industrial processes cannot be realized if the

L. B. Datsevich, *Conventional Three-Phase Fixed-Bed Technologies*,
SpringerBriefs in Applied Sciences and Technology,
DOI: 10.1007/978-1-4614-4836-5_7, © The Author(s) 2012

pressure drop exceeds 20–25 bar, the further rise of the specific productivity is impossible because of hydraulic limitations.

Even if the recycle compressors with a pressure boost greater than 20–25 bar were commercially available and the catalyst bed could endure the crashing impact due to the increased pressure drop, the productivity enhancement would demand a quadratic growth in the energy consumption, which is beyond any reasonable process efficiency.

The increased pressure in the reactor can facilitate the catalyst productivity via $C_{A,\text{inv}}$ (see Eq. (5.31)), but its influence can be insignificant for many industrial applications.

Hydrogenation of 3,4-dichloronitrobenzene to 3, 4-dichloroaniline can illustrate this point. Taking into account that $\beta_{A,s}$ is independent of the pressure, the specific productivity as a function of pressure in the reactor can be evaluated according to Eqs. (5.31) and (5.20). Figure 5.4 shows that in the wide diapason of the pressure from 100 to 300 bar, the catalyst productivity varies insignificantly. Namely, a 1.5-fold rise in the reactor pressure from its operation point (200 bar) to 300 bar leads to a growth of the catalyst productivity only by 5 %. Such a feeble growth cannot be regarded as an effective and efficient way for process intensification, especially, if the disproportionate costs of pipework, heat exchangers, and other loop elements at higher pressure are taken into account.

So far, the case of an active catalyst has been discussed in this chapter. The same conclusion with regard to the process intensification can be made if the overall reaction rate is determined not by external mass transfer, but by the catalyst activity, i.e., there is no concentration gradients of the reactants outside the catalyst particle, but at least the concentration profile of one of reactants can exist in the catalyst. In this case, the process could be intensified if the catalyst particles of a smaller size were loaded in the reactor.

It can be easily shown that if the effectiveness factor $\eta < <1$, the specific productivity of the catalyst bed is dependent on the particle diameter as

$$S.P. \sim \frac{1}{d_{\text{cat}}} \qquad (7.1)$$

Unfortunately, again, the enhancement of a reactor productivity cannot be realized in the traditional fixed-bed reactors, because it necessitates the proportional growth of the molar flow rate of the recycled gas associated with the concomitant square-law rise in the pressure drop over the entire loop system.

Thereby, we can conclude that the conventional technologies with gas recycling cannot be intensified. From the first stage of their design, their productivity is predetermined and restricted by the hydraulic relations dictated by thermodynamics of gas and gas–liquid flow. Together, with the extremely low efficiency of the energy utilization for process purposes, the absence of any potential for the process improvement can be declared.

In terms of Fig. 1.2, it means that the horse-drawn car cannot move two times quicker if the horse already runs at the limit of its strength.

In the light of this analysis, it becomes clear why any idea about the introduction of monolith catalysts or the so-called structured trickle-bed reactors (e.g., catalyst particles packed in the channels of monolith blocks [1]) is pointless from a thermodynamic point of view. This conclusion is more conclusive if operation and maintenance of comparatively big reactors is considered with regard to the procedures of loading and unloading, sealing the gaps (to avoid gas and liquid bypasses) between the monoliths themselves and monoliths and the reactor wall.

Since the significant part of experimental and theoretical efforts to enhance the existing trickle-bed reactors falls at the unsteady-state operation (e.g., feed modulation), the next chapter in detail discusses why these attempts are fruitless.

Reference

1. T. Bauer, S. Haase, Comparison of structured trickle-bed and monolithic reactors in Pd-catalyzed hydrogenation of alpha-methylstyrene. Chem. Eng. J. **169**(1–3), 263–269 (2011)

Chapter 8
Unsteady-State Operation (Feed Modulations) as an Attempt to Intensify Processes: Can it be Applied on a Great Scale?

As stated in the previous chapter, the traditional fixed-bed technologies with gas recirculation do not possess any potential for the process intensification.

Nevertheless, during the past two decades, many attempts have been undertaken to explore a possibility of enhancing the catalyst productivity in TBRs. Let us analyze some of these efforts in order to ascertain once more the lack of substantial prospects toward the process improvement.

An overwhelming number of such efforts falls at the unsteady-state mode of operation that is usually realized by the time modulation of the liquid feed by varying the volumetric flow rate (e.g. "on–off" feed strategy) [1, 2].

It is supposed that if the reaction is limited by gas–liquid mass transfer, the reactor should demonstrate better performance when the liquid is periodically pushed through the bed while the gas phase is fed incessantly. The reaction should be enhanced by the liquid flush that removes heat and products from the catalyst while the gas reactant can more easily penetrate into the catalyst bulk [1].

In experiments with the "on–off" feed modulation, the influence of two parameters on the reaction performance is investigated: period of modulation τ_Σ and time of feed τ_1, during which the liquid reacting compound A is fed into the reactor. Usually, the ratio $\psi = \tau_1/\tau_\Sigma$ is termed split. Typical values of the modulation period and split chosen in experiments lie in the range of 1–16 min and 0.25–0.4, respectively [3–6]. This means in order to keep the average liquid feed during the operation equal to $Q_{l,\text{feed}}$, the peak flow rate during feed time τ_1 should be equal to $Q_{l,\text{feed}}^{\text{peak}} = Q_{l,\text{feed}}/\psi$, which is 2.5–4 times greater than $Q_{l,\text{feed}}$. As is reported [4–6], the enhancement of the catalyst productivity by 10–30 % can be demonstrated in the course of such experiments.

Interpreting the experimental results, the supporters of the unsteady-state operation declare the possibility of its application to exothermic reactions on an industrial scale, e.g., in hydrogenation processes.

L. B. Datsevich, *Conventional Three-Phase Fixed-Bed Technologies*,
SpringerBriefs in Applied Sciences and Technology,
DOI: 10.1007/978-1-4614-4836-5_8, © The Author(s) 2012

Since there is not a single work that discusses how the feed modulation could be realized in the real industrial reactor with a gas loop where an exothermic reaction is carried out, let us propose some ideas about it.

Two ultimate cases attributed to the specifications posed on the hydraulics of the gas loop, as discussed in Chap. 5, can be supposed. The first case is related to the hydraulics adjusted to the nominal flow rate of the liquid reactant $Q_{l,\text{feed}}$, and the second is related to the case when the recycle compressor and gas loop allow for pushing the increased amount of gas corresponding to $Q_{l,\text{feed}}^{\text{peak}}$, which, in turn, is 2.5–4 times greater than $Q_{l,\text{feed}}$.

Could the second case be realized, any feed modulation was not necessary at all because according to Eq. (5.32) the reactor productivity should grow by 58–212 %, which is far higher than 10–30 % reported by supporters of the unsteady-state operation.

In the first case, when the reactor system has no margin with regard to recycled amount of gas, i.e., the molar flow rate of recycled gas $N_{g,\text{recycle}}$ is rigidly adjusted to $Q_{l,\text{feed}}$, the realization of the feed modulation can cause an irreversible effect with regard to the process selectivity and, what is more important, to the process safety.

As discussed in Sect. 5.8, even a slight increase in the flow rate of the liquid reactant $Q_{l,\text{feed}}$ can provoke the temperature excursion leading to the reactor runaway, let alone its rise by 2.5–4 times. Actually, if the flow rate $Q_{l,\text{feed}}$ is increased, almost the proportional growth of the adiabatic temperature rise of the reaction mixture ΔT can be expected. Moreover, the pulsing liquid inevitably increases the pressure drop in the part of the pipework, where two-phase flow is brought about, which additionally worsens gas recirculation and, hence, the heat removal.

The great menace to the process safety can arise not only during the time τ_1, when there is an increased flow of the liquid reactant A, but also during the time, when there is no feed. In the part of the catalyst bed near to its inlet, the catalyst particles are permeated by a highly concentrated liquid reactant. When the liquid flow is off, the liquid film around the catalyst becomes so thin that the oscillatory behavior of higher intensity accelerates the reaction rate (Sect. 5.3), and therefore the heat evolution. After the liquid is partially evaporated, the extremely severe and rapid temperature excursion under such conditions can occur. This can result in the reactor destructions, especially, if uneven gas distribution takes place. The discussion of such a scenario and its experimental verification can be found somewhere [7].

Thus, the feed modulation in any form can never be accepted by virtue of the safety precautions discussed already in Sect. 5.8.

Apparently, some comments on unsteady-state experiments should be given here. Namely, why no heat accidents have been encountered in experiments cited above even when the adiabatic reactors are used.

Purposefully or intuitively, the authors of these papers carried out their unsteady-state reactions far away from the conditions relevant to a real chemical production. In order to evade overheating the reaction mixture, and therefore degrading the selectivity, they used diluted feeds (e.g., 10-fold!), and did not achieve complete conversion. It is important to underline that both these conditions can never be adopted by the industry.

Actually, because of economic reasons, any chemical reactor including TBRs and BCRs should possibly process highly concentrated reactants.

As is well known, one of the main items of expense in any chemical production is the separation stage, in the course of which the ultimate product has to be extracted from the liquid mixture leaving a multiphase reactor. This stage of the chemical manufacture, as a rule, involves such energy consuming processes as distillation or crystallization accompanied with the following cost-intensive regeneration of used solvents. Thereby, the liquid feed should be of the high concentration of the initial compound A, and its conversion should be possibly full.

With regard to low operating costs, there is a common rule applied to chemical production (see examples in Tables 3.2 and 5.2): The feed should represent the initial compound A without any solvent if it is a liquid or can be in the liquid state under process temperatures. If the starting compound A is a solid or represents a very viscous substance, an appropriate solvent then is utilized, but its concentration should again be possibly high.

It is worth emphasizing once more that the high concentrations of liquid reactants are responsible for the high adiabatic temperature rises in several hundred and even thousand degrees (Table 3.2), which can be reached if the gas recirculation fails to withdraw the reaction heat. This point should never be ignored if an unsteady-state operation is considered.

Thus, unsteady-state operation cannot be regarded as a method for process improvement. It is likely a ground as to why so far there are no reports about industrial applications of unsteady-state operation on an industrial scale although the discussion of this topic goes on.

For fairness sake, it should be pointed out that for nonexothermic reactions the realization of which does not request the gas recirculation, the unsteady-state mode of operation may be useful to some extent.

References

1. K.D.P. Nigam, F. Larachi, Process intensification in trickle-bed reactors. Chem. Eng. Sci. **60**, 5880–5894 (2005)
2. P.L. Silveston, J. Hanika, Challenges for the periodic operation of trickle-bed catalytic reactors. Chem. Eng. Sci. **57**, 3373–3385 (2002)
3. M.I. Urseanu, J.G. Boelhouwer, H.J.M. Bosman, J.C. Schroijen, Induced pulse operation of high-pressure trickle bed reactors with organic liquids: hydrodynamics and reaction study. Chem. Eng. Sci. **43**, 1411–1416 (2004)
4. V. Tukac, M. Simickova, V. Chyba, J. Lederer, J. Kolena, J. Hanika, V. Jiricny, V. Stanek, P. Stavarek, The behaviour of pilot trickle-bed reactor under periodic operation. Chem. Eng. Sci. **62**, 4891–4895 (2007)

5. J. Hanika, V. Jiricny, P. Karnetova, J. Kolena, J. Lederer, D. Skala, V. Stanek, V. Tukac, Trickle bed reactor operation under forced liquid feed modulation. Chem. Ind. Chem. Eng. Q. **13**(4), 192–196 (2007)
6. G. Lia, X. Zhang, L. Wang, S. Zhang, Z. Mi, Unsteady-state operation of trickle-bed reactor for dicyclopentadiene hydrogenation. Chem. Eng. Sci. **63**, 4991–5002 (2008)
7. A.E. Kronberg, How to prevent runaways in trickle-bed reactors for pygas hydrogenation. Chem. End. Technol. **25**, 595–601 (2002)

Chapter 9
Alternative Industrial Fixed-Bed Technologies

As is well known, the former USSR kept the military parity with the developed counties, but was incapable to manufacture some chemical equipment, e.g., high-pressure compressors for hydrogen recirculation. Such compressors were employed, for instance, in the production of propellants, jet fuel additives, fiber monomers, pharmaceuticals, pesticides, and other intermediate and end chemicals for civil and military use. The cores of these manufacturing processes were the conventional three-phase fixed-bed reactors operated under high pressure. Because of the embargo imposed on the import of hydrogen recirculation compressors into the Soviet Union, the operation and maintenance of many industrial plants as well as the construction of new ones were problematic if possible at all.

The necessity to overcome the difficulties associated with the gas recirculation led to the development of two alternative techniques, the GIPKh and POLF technologies, which did not demand the gas loop and, therefore, circulating gas compressors.

The first technology (GIPKh is the Russian abbreviation for The State Institute of Applied Chemistry) represents a reactor with the gravitational film flow over the surface of a fixed-bed catalyst [1–4], from which the reaction heat is withdrawn by the liquid loop.

Up to now, the people, who were not knowledgeable about the fundamentals described in the foregoing chapters, are bewildered by the spectacular results demonstrated by this technology. Actually, the GIPKh technology is characterized by the same or greater catalyst productivity and higher selectivity under the pressure, which is several times less than that in the classical fixed-bed reactors with a gas loop. The more surprising fact is that such productivity is achieved not only under far less operating pressures, but also at much lower concentrations of liquid reactants entering the reactor.

Furthermore, the lower pressure and the less concentration of the liquid reactant ensure the exceptionally safe operation. The spontaneous runaways inherent in the traditional fixed-bed reactors (Sect. 5.8) can never occur in the GIPKh processes.

L. B. Datsevich, *Conventional Three-Phase Fixed-Bed Technologies*,
SpringerBriefs in Applied Sciences and Technology,
DOI: 10.1007/978-1-4614-4836-5_9, © The Author(s) 2012

It is worth underlining that the technical embodiments of the GIPKh processes are very simple, which sufficed at that time not only to construct new plants, but also to upgrade old fixed-bed reactors so that the expansive and cumbersome stage of gas recycling was excluded from operation. Afterwards, these old reactors revamped according to the GIPKh method ran under several times less pressures without a loss to the production output, product quality, and catalyst lifecycle.

The POLF technology (POLF stands for Presaturated One-Liquid Flow) represents the further development in multiphase techniques [1, 5–9]. The specific characteristic of this approach is the absence of the gaseous phase (i.e. as a separate phase) in the reactor at all. Only the liquid is directed into the catalyst bed, but this single liquid phase is previously saturated with gas outside the reactor in a specially designed device.

The main advantages of the POLF technology are associated with the extremely high catalyst productivity, far simpler equipment and absolute safety. The POLF technology demonstrates the catalyst productivity, which is ten and more times higher (in some cases a 100-fold growth can be achieved!) than that of the traditional TBR and BCR, whereas the operating pressure in the POLF system is several times less.

It should be mentioned that small catalyst particles (less than 1 mm) can be used in the POLF reactors. In this case, the overall reaction rate related to the catalyst mass of such small particles becomes comparable with the productivity of a suspended catalyst in slurry reactors (see Table 2.1). This offers a great challenge for the miniaturization of industrial reactors even of huge productive capacity.

As well as the GIPKH process, the POLF technique can be utilized for revamping the existing units to retrench the operating costs, improve product quality and safety, and enhance the catalyst productivity.

From a thermodynamic point-of-view, the GIPKh and POLF technologies are distinguished from the traditional approaches by the method for the energy supply. For the dissipative processes associated with mass and heat transfer in the catalyst bed, the energy is delivered by the moving liquid, not by the compressed gas. That makes the efficiency of the energy utilization up to 1000 times better if compared to the conventional fixed-bed technologies, and, therefore, results in a substantial decrease in the operating costs.

9.1 GIPKh Technology

9.1.1 Process Flow Diagram and Principal Embodiments

Two configurations of the GIPKh technology can be utilized in industrial applications—the one-stage and two-stage processes. In the one-stage process (Fig. 9.1), the whole volume of a catalyst is encompassed by the liquid loop. In the two-stage process, there are two catalyst sections situated above and beneath

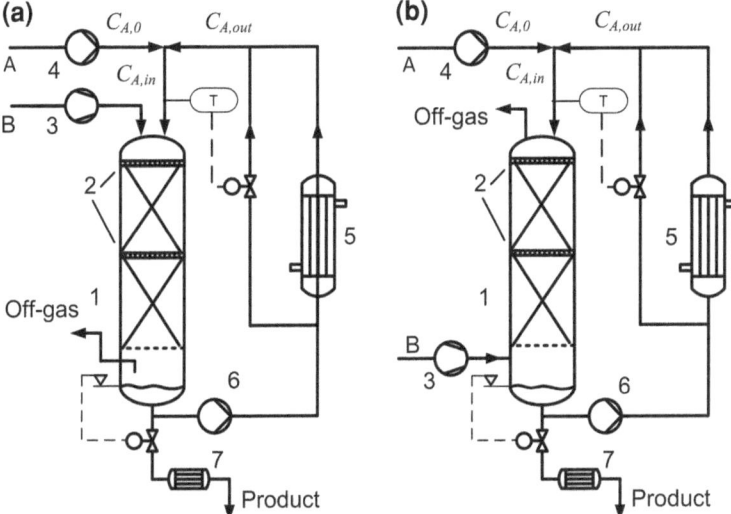

Fig. 9.1 Simplified technological schemes of the one-stage GIPKh process. **a** Concurrent, **b** Countercurrent. *A* liquid reactant, *B* gaseous reactant, *1* Reactor (with an embedded phase separator), *2* Gas–liquid (re)distributors, *3* Compressor for fresh gas, *4* Feed pump, *5* Loop heat exchanger (cooler), *6* Recycle pump, *7* Product cooler

intermediate phase separator 8 (Fig. 9.2). The upper section is involved in the liquid recirculation whereas the section below is excluded from it. In both systems, the liquid in the reactors is driven by gravity downwards.

In the one-stage GIPKh technology, liquid reactant *A* is mixed with the product, which is taken from the reactor outlet by recycle pump 6. The mixture of the liquid reactant and product (as well as a solvent if it is used) is directed to the catalyst bed where it is uniformly spread by means of gas–liquid distributor 2. Due to capillarity, the mobile liquid film is formed around the catalyst particles, ensuring efficient gas–liquid and liquid–solid mass transfer of both gas and liquid reactants to the catalyst surface.

If the height of the catalyst bed in reactor 1 is considerable, the installation of additional gas–liquid redistributors each 2–3 meters downstream can be demanded in order to facilitate the uniform flow along the whole catalyst volume.

Going through the catalyst bed, the liquid reactant is converted to the liquid product. This product is collected in the reactor bottom, which is used as a phase separator. From it, one part of the product is taken away by recycle pump 6 and returned to the reactor inlet through heat exchanger 5. Another part of the product is withdrawn from the reactor unit through cooler 7 with the help of the control system, whose task is to keep the constant liquid level in the phase separator.

Depending on the nature of the liquid feed and product, concurrent or countercurrent modes (Figs. 9.1a and 9.1b, respectively) can be realized. Compressor 3 introduces gaseous compound *B* either into the top of the reactor (Fig. 9.1a) or into its bottom (Fig. 9.1b).

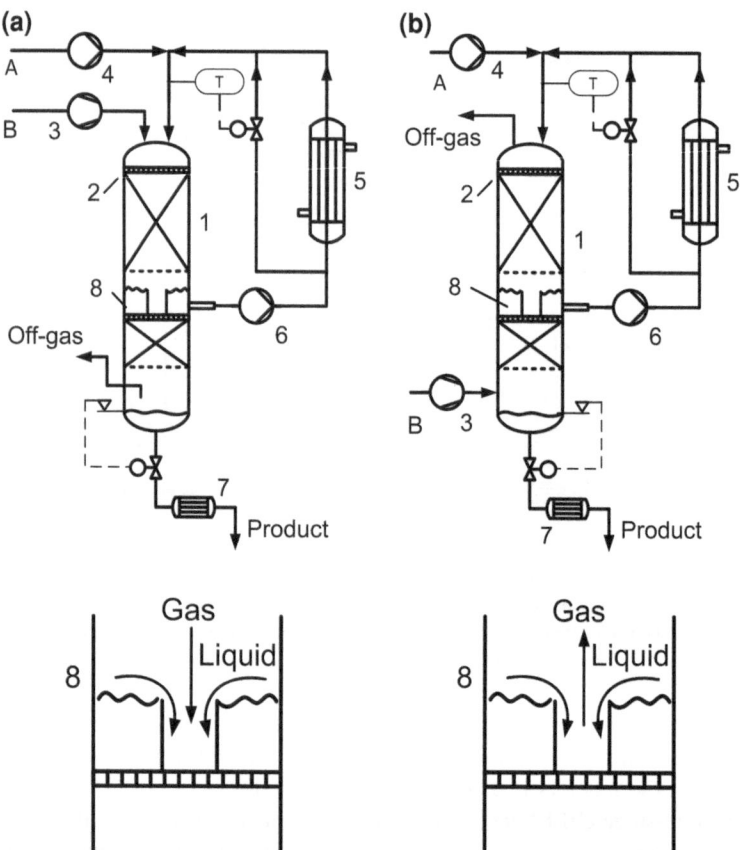

Fig. 9.2 Simplified technological schemes of the two-stage GIPKh process. **a** Concurrent, **b** Countercurrent. *A* liquid reactant, *B* gaseous reactant. *1* Reactor (with two embedded phase separators), *2* Gas–liquid distributor, *3* Compressor for fresh gas, *4* Feed pump, *5* Loop heat exchanger (cooler), *6* Recycle pump, *7* Product cooler, *8* Intermediate phase separator with an attached gas–liquid distributor below

The countercurrent mode can be advantageous when the starting compound is more volatile than the product. Due to partial distillation of the starting material by the counter flow of gas downstream and its condensation upstream, the higher conversion and more preferable temperature profile can be achieved.

The requested temperature of the recycled liquid at the reactor inlet T_0 is regulated by the control system including heat exchanger (cooler) 5 where heat generated in the catalyst bed is removed.

There are different methods for this temperature control. For example, the temperature can be governed by the flow rate of the coolant passing the heat exchanger. However, such a method is rather seldom applied to industrial GIPKh plants because of its exceptional inertia.

For a fast, nearly non-inertial response, another method is more appropriate. The hot product taken from reactor 1 by recycle pump 6 is divided into two parts. One part goes through heat exchanger 5 where it cooled below the requested temperature at the reactor inlet. The second part bypasses cooler 5 and is mixed with the first, chilled stream so that temperature at the reactor inlet T_0 can be immediately adjusted to the demanded value by the regulating valve installed, for example, in the bypass line.

The two-stage GIPKh process (Fig. 9.2) is applied when the extremely high conversion is demanded. In this reactor, the upper catalyst section functions similarly to the one-stage GIPKh reactor. Intermediate phase separator 8, which is supplemented with a gas–liquid distributor below, automatically splits the liquid flux into two parts. The first one is returned by recycle pump 6 to the reactor inlet while the second one flows over the phase separator opening, goes through the attached gas–liquid distributor and irrigates the catalyst bed underneath. In the second catalyst section, the ultimate conversion of liquid reactant A can be achieved. The product is collected in the bottom of the reactor and withdrawn through cooler 7 by the level control system.

The pressure in all the GIPKh schemes (Figs. 9.1 and 9.2) is controlled automatically with the help of compressor 3. In order to prevent the accumulation of gaseous by-products, the continuous ventilation of the reactor should be foreseen. Usually, the purging flow rate about 3–5 % of the chemical consumption can guarantee the absence of such a situation. In some processes, for example, in hydrogenation of furfural to furfuryl alcohol, the increased ventilation can be required during the first several days after a freshly activated Cu catalyst begins its cycle. During this time, the extremely active catalyst partly decomposes furfural into furan and carbon monoxide. The latter will gather in the reactor fading the reaction if it is not withdrawn from the system.

Recycle pump 6 in Figs. 9.1 and 9.2 is a key element of the system accounting for its entire functioning. Unsophisticated hermetic single-stage centrifugal pumps of low-pressure head (up to 50 m), but elevated working pressure (50–100 bar) can be utilized. The main advantage of such pumps is their compact design, reliable operation, reasonable price, and low maintenance requirements.

The number of the different GIPKh configurations is not covered exclusively by Figs. 9.1 and 9.2. Each phase separator can be represented by a separate apparatus; the two-stage process can be carried out not in the single reactor, but in the train of two (or more) reactors.

The same commercial catalysts as used in the conventional fixed-bed reactors can be exploited in the GIPKh processes.

Catalysts with metal oxides (e.g. Ni, Cu and Co catalysts) can demand the activation prior to their operation cycle. This procedure does not differ from the reduction process in the traditional fixed-bed reactors with gas recirculation described already in Chap. 4. The corresponding equipment functioned under atmospheric pressure should additionally be installed for this purpose (not shown in Figs. 9.1 and 9.2).

Other types of the commercial catalysts described in Chap. 4 can be activated in the framework of the apparatuses indicated in Figs. 9.1 and 9.2.

9.1.2 Temperature Control and Heat Balance

Two operating parameters define the temperature profile along the catalyst bed: temperature at the reactor inlet T_0 and recirculation rate of the liquid phase K_{recycle}. The latter predetermines adiabatic temperature rise of the reaction mixture ΔT.

In the two-stage process, the conversion of compound A after the first catalyst section should be taken into account with regard to the further adiabatic temperature growth in the second catalyst section.

The adjustment of a temperature profile in the GIPKh reactors follows the same strategy as in the conventional fixed-bed systems (Fig. 5.1). When the catalyst loses its activity, the temperature of the liquid mixture at the reactor inlet T_0 will be increased while the admissible temperature rise ΔT_{permit} is retained constant during the whole operation cycle.

Contrary to the technologies with a gas loop where the reaction heat is absorbed by the tremendous amount of gas passing through the catalyst bed, in the GIPKh processes, the gas has practically no impact on temperature. Demanded temperature rise ΔT_{permit} in the reactor is ensured by the recirculation of the liquid phase.

For reactors with diameters more than 10–20 cm, heat losses through the reactor wall can be neglected. Heat generated in the course of the reaction is directed at warming the liquid mixture by temperature ΔT_{permit} in accordance with Eq. (9.1):

$$C_{A,0}Q_{l,\text{feed}}X(-\Delta H_A) = \rho_l(Q_{l,\text{feed}} + Q_{\text{recycle}})C_{P,l}\Delta T_{\text{permit}} \tag{9.1}$$

Introducing recirculation rate K_{recycle} as

$$K_{\text{recycle}} = \frac{Q_{\text{recycle}}}{Q_{l,\text{feed}}}, \tag{9.2}$$

one can obtain from Eq. (9.1)

$$K_{\text{recycle}} = \frac{C_{A,0}(-\Delta H_A)X}{\rho_l C_{P,l}\Delta T_{\text{permit}}} - 1 = \frac{\Delta T_{\text{ad}}}{\Delta T_{\text{permit}}}X - 1 \tag{9.3}$$

Here $\Delta T_{\text{ad}} = C_{A,0}(-\Delta H_A)/(\rho_l C_{P,l})$ is the maximum adiabatic temperature rise in the absence of liquid recirculation, i.e., at $Q_{\text{recycle}} = 0$ [compare it with Eqs. (5.1) and (5.3)].

Adiabatic temperature rise ΔT in the catalyst bed involved in the liquid recirculation can be estimated from Eq. (9.3) under assumption of complete conversion of compound A (i.e. $X = 1$),

$$\Delta T = \frac{\Delta T_{ad}}{K_{recycle} + 1} \tag{9.4}$$

Equations (9.1), (9.3), and (9.4) reflect a very simple idea: dilution of initial compound A by the liquid product proportionally reduces the adiabatic temperature rise. Thus, adjusting appropriate recirculation rate $K_{recycle}$, it becomes possible to control the temperature in the catalyst bed without a gas loop in the range preconditioned by the technological reasons, viz.: between T_0 and $T_0 + \Delta T_{permit}$.

9.1.3 Why were the Reactors with a Liquid Loop Not Considered for Industrial Applications Earlier?

Despite the apparent simplicity of the temperature control by means of liquid recirculation, this method was never regarded as an alternative to the conventional techniques, at least down to recent times. Some reasons (myths) lying in this fact have already been elucidated in Chaps. 3 and 5.

The last objection raised by supporters of the old technologies is discussed in this Chapter. Their argument is related to the particularity inherent in the GIPKh (and POLF) reactors, which is characterized, on the one hand, by the considerably lower concentration of liquid reactant A in the catalyst bed in comparison to the traditional reactors and, on the other hand, by the incredibly high flow rate of the liquid phase through the catalyst bed. (In other words, from the point-of-view of the residence time distribution, the flow pattern of the GIPKh (and POLF) reactors at high recirculation rates is similar to the continuously stirred-tank reactor (CSTR).)

Actually, proceeding from ΔT_{ad} and ΔT_{permit} given in Table 3.2, one can find recirculation rate $K_{recycle}$ (Eq. 9.3). In some cases, it can be more than 100. Correspondingly, this entails a diminution in the concentration of compound A at the reactor entrance. Denominating this concentration as $C_{A,in}$ and writing the mass balance at the reactor inlet as

$$Q_{l,feed}C_{A,0} + Q_{recycle}C_{A,out} = C_{A,in}(Q_{l,feed} + Q_{recycle}), \tag{9.5}$$

one can obtain

$$C_{A,in} = \frac{C_{A,0} + K_{recycle}C_{A,out}}{1 + K_{recycle}} = C_{A,0}\frac{1 + K_{recycle}(1 - X)}{1 + K_{recycle}} \tag{9.6}$$

As can be seen, the inlet concentration of liquid reactant A in the GIPKh processes $(C_{A,in})$ is only a small fraction of that in the conventional fixed-bed reactors $(C_{A,0})$ while the rate of the liquid flow through the catalyst bed is much higher.

When the idea about such a method for temperature control was first proposed, there were great doubts about possibility of reaching the catalyst productivity acceptable for industrial purposes. It was expected that the much lower concentration of the liquid reactant together with the increased flow rate should lead to a considerable growth of the reactor size.

This argument against the GIPKh technology can be illustrated with hydrogenation of furfural to furfuryl alcohol.

For required conversion $X = 0.99$ in the industrial BCR with hydrogen recirculation, the operating pressure is 150 bar, and 100 % furfural enters the catalyst bed ($C_{A,0} = 12.1 \times 10^3$ mol/m^3). In the GIPKh process at given $\Delta T_{ad} = 1200°C$ and $\Delta T_{permit} = 20°C$ (see Table 3.2), $K_{recycle} = 58.4$ and the concentration of furfural entering the catalyst bed is 2.7 % ($C_{A,in} = 322$ mol/m^3), which is 37 times less if compared to BCR.

Omitting simple mathematical manipulations, the ratio between the catalyst volumes in the GIPKh and traditional reactors at the same hydrogen pressure can be calculated as

$$\frac{V_{bed,GIPKh}}{V_{bed,conventional}} = \frac{(1 + K_{recycle}) \ln\left(\left(\frac{1}{1-X} + K_{recycle}\right)/\left(1 + K_{recycle}\right)\right)}{\ln\left(\frac{1}{1-X}\right)} \tag{9.7}$$

Equation (9.7) is derived under assumptions of the plug flow behavior and the first-order effective (formal) kinetics with regard to liquid reactant A in both reactors.

According to Eq. (9.7), the GIPKh reactor should have the catalyst volume about eight times more than the traditional BCRs or TBRs with a gas loop, which can never be accepted for industrial applications.

It is appropriate to mention here that the industrial BCR reactor for the reaction considered in the example above was upgraded according to the GIPKh process (Fig. 9.1). Although many professionals with outstanding achievements in developing TBR and BCR with a gas loop had predicted an inevitable fiasco, the revamped reactor demonstrated the same productivity, better selectivity, and longer catalyst life not at the pressure of 150 bar, but three times less—at 50 bar.

9.1.4 Hydrodynamic and Mass Transfer Aspects of the GIPKh Reactors

In the GIPKh reactors, the gravitational film flow over the surface of randomly oriented catalyst particles is realized. The film should be comparatively thin in order to afford the free motion of the gas phase in the channels between the catalyst particles necessary for sufficient mass transfer.

Under conditions of the laminar flow and negligible shear stress at the gas–liquid interface, film thickness δ on a vertical wall can be derived from the Navier–Stokes equation

$$\delta = \left(\frac{3 U_l d_{\text{cat}} \eta_l}{4(1 - \varepsilon) \rho_l g} \right)^{\frac{1}{3}} \quad (9.8)$$

Unfortunately, Eq. (9.8) is applicable only to a short part of the real film on the catalyst surface somewhere beyond the contact points with other particles. In such points, the liquid forms some kind of droplets, the size of which depends on the physical and geometrical properties (e.g. surface tension, density, contact angle, etc.).

Although superficial velocity of liquid U_l has a slight impact on the film thickness ($\delta \sim \sqrt[3]{U_l}$), it should always be the focal issue during design procedures. At elevated liquid velocities, the gas channels can be sealed due to thickening the film, which, in turn, can lead to the flooding of the catalyst bed. If the flooding occurs in some part of the catalyst bed, the reactor will not function more. In order to evade the catalyst flooding, superficial velocity of liquid U_l should never exceed its critical value during the operation.

With regard to the flooding, the particular attention should be paid to the countercurrent flow of gas and liquid (Figs. 9.1b and 9.2b). By virtue of the shear stress applied to the gas–liquid interface in the upward direction, the liquid film becomes thicker by contrast to the concurrent mode. Since the gas velocity of gas varies from its maximum at the entrance into the catalyst bed to practically zero at its top, the bottom of the catalyst bed can be the most problematic place.

The relation between the maximum of superficial gas velocity U_g and superficial velocity of liquid U_l can be expressed as follows

$$U_g = n C_{A,0} \frac{RT}{P} \frac{1}{1 + K_{\text{recycle}}} U_l \quad (9.9)$$

Equation (9.9) does not reflect an increase in the gas velocity due to partial evaporation of liquid into the dried reacting gas at its entrance into the catalyst bed. This rise in the velocity should always be figured out during designing the industrial reactors.

In countercurrent mode, the boundary between the film flow and the flooding can be expressed as a function $U_l = f(U_g)$, in which higher U_l corresponds to lower U_g. In reactors with catalyst particles of 3–6 mm, liquid velocity U_l should be not more than 0.8 cm/s; gas velocity U_g should be limited to 4–6 cm/s.

It is obvious that the film flow can pose some restrictions on the reactor geometry. If the feed flow rate is preset, it can happen that the GIPKh reactor should be of a disproportionate diameter, which cannot be accepted for the apparatus operated under pressure.

Under liquid velocities $U_l \sim 0.5 - 0.8$ cm/s, film thickness δ according to Eq. (9.8) is about 0.15–0.18 mm. If the average film thickness is calculated on the basis of liquid hold-up (i.e. taking into account the liquid gathered in sites of particle contacts), its value can be about 0.5 mm.

Under the real reaction conditions, the film performance can significantly differ from the flow pattern observed in the inert hydrodynamic experiments.

In exothermic reactions, as has already explained in Sect. 5.3, the oscillatory motion of liquid can occur in catalyst pores and sites of particle contacts. On the other hand, a temperature gradient across the liquid film reaches values of 10–20 °C/mm, which can trigger the Marangoni convection.

The oscillatory motion of liquid in the direction normal to the film boundary and the Marangoni effect should accelerate both gas–liquid and liquid–solid mass transfer. For that reason, the application of the existing mass transfer correlations for the film flow cannot be recommended for scaling-up since they are obtained in "inert" experiments and fail to predict the behavior of the "living" reacting system.

In industrial GIPKh reactors, the partial evaporation of liquid can lead to the drying of the catalyst bed. In hydrogenation of acetone to isopropanol at 10 bar, this effect was observed in an industrial reactor. In the countercurrent mode, the reduction of recirculation rate $K_{recycle}$ below some level dries a certain catalyst layer, which resulted in the fall of conversion.

9.1.5 Approximate Algorithm of the Research and Design Procedures

The definition of operating parameters and appropriate reactor dimensions with regard to either working out a new process or upgrading an existing fixed-bed reactor (TBR or BCR) is a key task of research and design procedures.

A small-scale working replica of a commercial GIPKh process operated in the continuous mode can deliver all necessary information pertaining to product quality, catalyst activity, and catalyst aging. Such essential data as operating pressure, and, if a new process should be worked out, permitted temperature rise ΔT_{permit} or, consequently, liquid recirculation rate $K_{recycle}$ can be specified in these experiments.

In the evaluation of the reactor size, the first step is a choice of operating superficial velocity of liquid U_l, on the basis of which the reactor diameter can be determined. It can be recommended to set this value by about 20 % less than the velocity of flooding. The second step is to ascertain the length of a catalyst bed necessary for demanded conversion. As a rule, mathematical modeling cannot be applied with confidence for this purpose because of uncertainties in mass transfer. As has been mentioned in Sects. 5.3 and 9.1.4, the traditional correlations can underestimate the rates of mass transfer processes.

In [1], the unsophisticated method for direct physical modeling is reported as a reliable option to the mathematical simulation. By the consecutive runs of the reaction mixture through a comparatively small experimental reactor, the real concentration profile of liquid reactant A in the future commercial unit as a function of the catalyst length can be obtained in a relatively short time and, as a result, the height of the catalyst bed for requested conversion can surely be estimated.

9.1.6 Comparison of the GIPKh Technology with the Conventional Processes

A couple of dozen industrial processes were worked out according to the GIPKh technology. About half of these projects were realized in the framework of revamping the traditional TBR or BCR with gas recirculation. These retrofitted reactors are of particular interest. Their efficiency, safety, and reliability were directly compared to the TBR and BCR since the reactors, catalysts, temperature regimes, quality of feeds and solvents, infrastructure on sites, and operating personnel had been the same before and after the retrofit.

The specific characteristics of some of these processes are presented in Table 9.1.

As is indicated, the same or higher catalyst productivity can be achieved at a three- to fivefold decrease in the operating pressure. That additionally confirms the correctness of the analysis related to the conventional technologies and dispels the myths about them (Chap. 5). It is also worth emphasizing that the GIPKh processes consume much less energy than the traditional methods.

As a rule, the industrial GIPKh processes also demonstrate the better selectivity. This fact is explained by the dilution of the high concentrated liquid feed with the product, which prevents the formation of the "hot spots" inherent in the traditional TBR and BCR (Sect. 5.8). The same reason lies in the exceptionally safe operation. The spontaneous temperature excursion without any detectable change in process parameters or a fast temperature jump due to any malfunction as it occurs in TBR and BCR is impossible. Even if the coarse control error happens, there will be enough time to shut down the process without any dangerous consequences.

The simple technical embodiments of the GIPKh technology are also of great advantage, which implies low investment costs for new construction and a comparatively quick revamp of the existing plants.

Unfortunately, there are also several drawbacks in the GIPKh processes. They are associated with the liquid flow, the velocity of which should never exceed its maximum because of flooding. Such rigid hydrodynamic conditions impose some restrictions on the reactor design if a comparatively large output is demanded. In this case, the computed reactor diameter can be incommensurable with its height, so that the installation of several parallel reactors should be considered. (This situation is comparable with the hydrodynamic restraints on the gas loop in the conventional TBRs and BCRs (Sect. 5.2), when several parallel reactors are often used).

Because of such a limit, the further process development in terms of Fig. 1.2 is impossible. As soon as this maximum velocity is achieved, the GIPKh technology will have no potential for enhancement.

Table 9.1 Some hydrogenation processes before and after revamp (after [1])

Process	Traditional fixed-bed technologies with gas recirculation (BCR and TBR)			GIPKh technology		
	Pressure (bar)	Specific productivity (h^{-1})	Energy consumption (kJ/mol_A)	Pressure (bar)	Specific productivity (h^{-1})	Energy consumption (kJ/mol_A)
Nitroparaffins (C12–C14) to aminoparaffins	150–200	0.15	400	50	0.15	4.9
3,4-dichlornitrobenzene to 3,4-dicloraniline	200	0.2	370	50	0.2	6.2
Furfural to tetrahydrofurfuryl alcohol	150	0.1	126	50	0.1	2.1
Furfural to furfuryl alcohol	100–150	0.15	43	50	0.15	0.7
Acetone to isopropanol	50	0.4	300	10	0.8	0.2

9.2 POLF Technology

Although the GIPKh technology excels the traditional techniques, it is incapable of utilizing the full potential of an active catalyst. By virtue of hydrodynamic restrictions inherent in the gravitational film flow, neither small catalyst particles can be used nor can external mass transfer be intensified. Nevertheless, the GIPKh technology sheds light on some essential points. For example, higher catalyst productivity can be achieved under much less operating pressure and extremely low concentration of the liquid reactants.

The POLF technology is the further development in multiphase fixed-bed reactors. This technique is characterized by extremely high productivity and safe standards, low energy consumption, and simple equipment. The main peculiarity of this method is in carrying out the ineffective gas–liquid mass transfer stage not in the reactor, but outside it in a specialized device.

9.2.1 Scientific Fundamentals of the POLF Technology

If all the fixed-bed technologies mentioned above are considered from a general point-of-view, one can draw a conclusion that the catalyst bed combines at least two different functions of both a chemical convertor and a mass transfer device. Without doubt, the catalyst is truly destined for the first one while, from the author's point-of-view, the second function is unsuitable for the catalyst bed, particularly, with respect to gas–liquid mass transfer.

As is followed from the analysis (Chap. 5 and Sect. 9.1), the execution of these two functions in one reactor is a rather difficult task if the attention is paid to hydrodynamic conditions.

In the POLF reactor, there is no gas–liquid mass transfer at all. This stage is carried out outside the reactor in a specialized and, therefore, highly effective device called "saturator".

Gas as a separate phase does not exist in the reactor (more exactly, no separate gas phase is observed). Only the liquid phase flows through the catalyst bed, but this liquid phase is previously saturated with gas outside the reactor in the saturator. As the single liquid is pumped through the catalyst bed, all the problems related to the pressure drop, energy consumption, and use of small catalyst particles are not so crucial as in the conventional TBR and BCR. The reactor productivity can be considerably enhanced by increasing the velocity of liquid or decreasing the size of catalyst particles.

In order to comprehend the POLF process, let us return to the two-zone model considered in Sect. 5.5.2. As has been already explained in Fig. 5.3, the catalyst bed in the conventional TBRs and BCRs can be thought of as consisting of two zones. The first zone is situated between the liquid inlet ($l_{cat} = 0$) and the inversion point ($l_{cat} = l_{inv}$) and the second one lies between the inversion point and the liquid

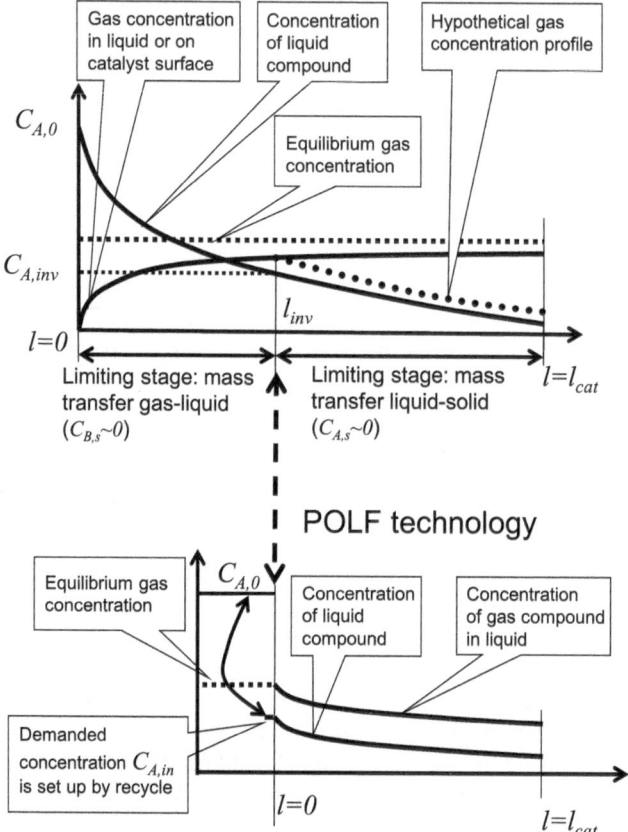

Fig. 9.3 Scientific principles of the POLF technology (after [1])

outlet ($l_{cat} = l_{bed}$). In the first zone, the size of which is much less than the length of the second one, the reaction rate is limited by gas–liquid mass transfer. In the second zone, the reaction rate is determined by liquid–solid mass transfer. At the point $l_{cat} = l_{inv}$, the proportion between gaseous and liquid reactants becomes equimolecular in terms of Eq. (3.1). Hypothetically, if beginning from this inversion point, the gas flow was shut down, no change in the reaction rate would be registered.

In the POLF reactor, this hypothetical second zone is exactly realized (Fig. 9.3). The liquid is introduced into the catalyst bed with the gas concentration $C_{B,l}^*$ corresponding to the gas–liquid equilibrium whereas the concentration of liquid compound A is taken equal to the equimolecular value $C_{A,in} = \frac{1}{n} C_{B,l}^*$. (Note: Because of technological reasons, the inlet concentration of liquid compound A

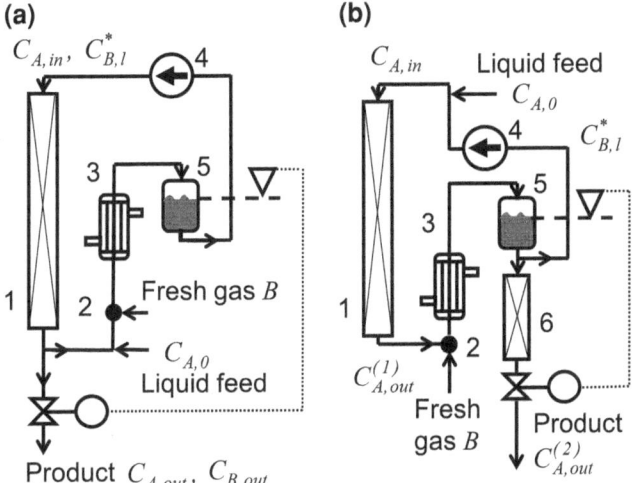

Fig. 9.4 Simplified flow diagrams of the POLF technology. One-stage (a) and two-stage (b) processes. *1* Reactor, *2* Saturator, *3* Cooler, *4* Recycle pump, *5* Phase separator, *6* Second reactor (after [6])

should be set less than this value (see Sect. 9.2.3).) Such a small concentration of liquid compound *A* is arranged through the recirculation of liquid by analogy with the GIPKh technology.

9.2.2 Process Flow Diagram and Principal Embodiments

The POLF technology is distinguished by the flexibility that allows one to adapt it to different industrial applications. In this Chapter, only some typical configurations—the one- and two-stage POLF processes are described in detail (Fig. 9.4). These processes are designated for carrying out the reactions, in which the concentration of liquid compound *A* in the feed is considerable if compared to purification processes, e.g., ultra-deep hydrodesulphurization.

In the one-stage POLF process (Fig. 9.4a), there is a single catalyst bed involved in the recirculation of the product. In the two-stage process (Fig. 9.4b), there are two catalyst beds, one of which is encompassed by the liquid recirculation and another is not.

In the one-stage process, the liquid feed with the concentration of liquid reactant *A* equal to $C_{A,0}$ is mixed with the product recirculated with the help of pump 4. This mixture is forwarded through saturator 2 into phase separator 5. In the saturator, the liquid is saturated with the reacting gas up to its maximal possible (equilibrium) concentration $C_{B,l}^*$. From phase separator 5, the liquid is taken by recycle pump 4 and led into reactor 1. At the reactor outlet, the product is divided into two streams. One is withdrawn from the POLF process by the controlling system keeping the constant level of liquid in phase separator 5, and another is recirculated by pump 4.

For extremely high productivity (or conversion), the two-stage processes (Fig. 9.4b) can be recommended. In this process, the product after the first reactor 1 is saturated with the gas in saturator 2 and split into two streams after phase separator 5. One part is recycled by pump 4, another part passes second reactor 6, where the ultimate conversion is achieved.

The temperature and pressure control in the POLF processes are similar to that in the GIPKh technology.

It is worth pointing out that the POLF reactors do not demand any liquid (let alone gas–liquid) redistributors. However, a splash plate may be needed at the entrance of liquid in the catalyst bed in order to protect the catalyst particles from the straight fluid jet.

The functionality of saturator 4 (in other words, carrying out gas–liquid mass transfer outside the catalyst bed) is based on the utilization of the kinetic energy of liquid, and can be realized, for example, in a simple static mixer. MPCP GmbH suggested a device that combined the recycle pump, saturator and phase separator in a single assembly [7, 10]. Such a device can be recommended for applications in the two-stage or multistage processes.

Depending on the specific features of each reaction, the saturator, phase separator and heat exchanger can be located somewhere in the liquid loop in the order differing from that in Fig. 9.4. Besides, two (Fig. 9.4b) or more catalyst beds can be placed in a single reactor, which simplifies the unit design.

As in the GIPKh technology, hermetic single-stage centrifugal pumps of low-pressure head and elevated working pressure can be used for liquid recirculation.

Since the single liquid phase is forwarded through the catalyst bed, the flow direction (downwards or upwards) has no influence on the process behavior. The reactor can be mounted in every possible orientation: horizontal, vertical, or any angle. Moreover, the gravity plays no role in its function (if MPCP device is used), which allows one to deploy the POLF system on board of a flying vehicle [7].

For purification processes, the POLF technology can be applied without a liquid loop in the single catalyst bed [7–9].

The POLF process may need the gas purging if a comparatively high amount of by-gases is formed in the course of the reaction. For this purpose, the controlled gas relief from phase separator 5 (not shown in Fig. 9.4) should be foreseen. It can also be provided with the help of MPCP device [7, 10].

It is necessary to mention that there are other modifications of the POLF technology, in which the gas concentration in the liquid phase can be kept the same throughout the catalyst bed. In them, the so-called "oversaturated" state of gas is used (Note: This term is not correct from a physical point-of-view because the gas concentration does not exceed its equilibrium value.) This state is characterized by the existence of small, invisible gas bubbles formed in the liquid phase owing to a powerful hydrodynamic impact of liquid on gas, for example, in the MPCP device [10]. Because of hydrodynamic conditions in the whole system, coalescence of such microbubbles can be prevented so that these microbubbles support the same equilibrium concentration of gas in the reactor. Moreover, the MPCP devices allow one to create a multistage process in one reactor, the residence time behavior of which is similar to the ideal plug flow pattern.

9.2.3 Choice of Process Variables

The choice of the operating parameters for the POLF technology represents a more difficult task than in the traditional technologies. In order not to overburden the text, this Chapter deals only with the process variables for the one-stage process according to Fig. 9.4a.

When the concentration of liquid reactant A in the feed and the product ($C_{A,0}$ and $C_{A,\text{out}}$, respectively) as well as maximum allowable temperature $T_{\max}$ preconditioned by the process selectivity, the rate of catalyst aging and safety are known, three operating parameters should be specified: operating pressure, conversion of dissolved gas B denominated as X_B and recirculation rate K_{recycle}.

The total pressure in the system, which is created by the reacting gas, defines the equilibrium concentration of the gas compound at the reactor inlet $C_{B,l}^*$. The elevated pressure facilitates the reactor productivity, but increases the costs of the applicable equipment. As a rule, the pressure several times less than in the conventional TBR and BCR enables, nevertheless, the reactor productivity several times more.

When the operation pressure is preset, the concentration profiles of the reactants, especially, of gas compound B along the catalyst length should be taken into account. According to Fig. 9.4a, the liquid enters the catalyst bed with the maximal concentrations of both reacting compounds A and B – $C_{A,in}$ and $C_{B,l}^*$, respectively. The latter corresponds to the gas–liquid equilibrium. When the liquid passes the catalyst, these concentrations will become less and reach their minimum at the reactor outlet- $C_{A,\text{out}}$ and $C_{B,\text{out}}$, respectively.

The excess of liquid reactant A or shortage of gas compound B should be avoided. Otherwise, if liquid reactant A is in excess (i.e. $C_{A,\text{in}} > \frac{1}{n} C_{B,l}^*$), gas compound B will be consumed completely somewhere in the catalyst bed so that some part of the catalyst downstream will not participate in the reaction. This leads not only to the less reactor productivity, but also to the faster catalyst deactivation. Therefore, it is recommended not to have the conversion of gas compound B

$$X_B = \left(1 - C_{B,\text{out}}/C_{B,l}^*\right) \text{ more than 0.8.}$$

Equilibrium gas concentration $C_{B,l}^*$ (that is determined by gas partial pressure and gas solubility) and requested gas conversion X_B determine recirculation rate K_{recycle}

$$K_{\text{recycle}} = \frac{n\left(C_{A,0} - C_{A,\text{out}}\right)}{X_B C_{B,l}^*} - 1 \tag{9.10}$$

Equation (9.10) can easily be derived from Eq. (9.5) (mass balances of the one-stage POLF and GIPKh processes are identical) by inserting $C_{A,\text{in}} - C_{A,\text{out}}$ through $X_B C_{B,l}^*/n = C_{A,\text{in}} - C_{A,\text{out}}$.

Temperature rise of the reaction mixture ΔT can be calculated over the recirculation rate

$$\Delta T = \frac{\Delta T_{\text{ad}}}{1 + K_{\text{recycle}}} \tag{9.11}$$

For an overwhelming number of processes under pressure about 50–100 bar, this temperature rise is less than 10–20 °C. In very rare situations (extremely high pressure or gas solubility), it can turn out that temperature rise ΔT calculated according to Eq. (9.11) is more than the permissible value ΔT_{permit}. In this case, the recirculation rate should be taken according to Eq. (9.3).

9.2.4 Phenomenology of Mono Phase Flow and Liquid Solid Mass Transfer

In comparison to the conventional and GIPKh technologies, the POLF technology has neither peculiarities nor uncertainties related to hydrodynamics and mass transfer. At least, there is no great discrepancy in mass transfer correlations obtained by different research groups [11]:

$$Sh_{i,s} = 1.17 \text{Re}_l^{0.585} Sc_{i,l}^{1/3} \tag{9.12}$$

Nevertheless, the considerations associated with the oscillations in catalyst pores remain in force. The attention should be paid to unpredictable acceleration of mass transfer.

If the catalyst of high productivity is used, the reaction rate under chosen process parameters (Sect. 9.2.3) will be limited by mass transfer of liquid reactant A. For this case, Eq. (9.12) shows the practically limitless way of the enhancement of the reactor productivity. A rise in the flow velocity through the catalyst bed results in an increase in mass transfer proportionally to $U_l^{0.585}$. A decrease in particle size leads to mass transfer enhancement by factor of $\beta_{i,s} a_s \sim \frac{1}{d_{\text{cat}}^{1.415}}$.

It is necessary to underline that both the high liquid velocity and the small catalyst particles have not such a hydrodynamic impact on the system as in the case of traditional TBRs and BCRs. In addition, not only is the energy consumption in the POLF system low, but its utilization is also very efficient: the energy dissipation occurs mainly in the catalyst bed where it is demanded.

9.2.5 Approximate Algorithm of the Research and Design Procedures

The research algorithm related to the product quality, catalyst aging, and temperature regime does not differ from the procedures described already for the conventional TBRs and BCRs or GIPKh processes.

Unlike the traditional TBRs and BCRs as well as the GIPKh reactors, the POLF technology allows for the use of small catalyst particles in size less than 1 mm. Although the particles of the same size (>1 mm) as in the traditional TBRs and BCRs already demonstrate much higher productivity in the POLF reactors under far less operating pressure, small particles result in still more compact design and, therefore, better process efficiency.

The enhancement effect of small catalyst particles if an active commercial catalyst is applied can be estimated for two limiting cases: limitations caused by (i) external mass transfer or (ii) intraparticle diffusion. In the first case, a decrease in a catalyst size enhances the reactor productivity or reduces the catalyst volume by factor of $d_{cat}^{-1.415}$. In the second case, this enhancement factor is proportional to d_{cat}^{-1} (Eq. (7.1)).

The pressure drop over the catalyst bed should be taken into consideration when the further process intensification by means of decreasing the size of catalyst particles is desired. In the first and second cases for the same reactor productivity and diameter (but less catalyst volume), the pressure drop grows proportionally to the factor of $d_{cat}^{-0.585}$ or d_{cat}^{-1}, respectively.

Since the design of the POLF systems is very compact (reactor of small volume and short-length pipework), the operating pressure is not so crucial for a choice of equipment. In many industrial processes, as a rule, the operating pressure below 50 bar can be sufficient and, therefore, low-pressure apparatuses can be used.

According to Sect. 9.2.3, the great attention should be paid to such physical properties as gas solubility. The solvent, in which the equilibrium concentration of the reacting gas is higher, is more preferable. For example, the solubility of hydrogen in different solvents can vary up to 15-fold.

A special structure of flow through the catalyst bed, for example, by sectionalizing the reactor room [1] can also increase the specific productivity of the catalyst. It can also be a purpose of the research.

Scale-up of POLF reactors can be carried out in any laboratory without difficulty. Any uncertainties with respect to kinetics and mass transfer can be avoided in such scale-up experiments. Since the orientation of the catalyst bed plays no role, the length of a future industrial reactor can be imitated by a number of small tubes with a catalyst connected one after another in a bundle [1], so that the concentration profiles can be obtained by sampling the liquid after each section.

Table 9.2 Comparison of the conventional TBRs/BCRs with the one-stage POLF process

Reaction	Conventional technology		POLF technology	
	Pressure (bar)	Specific productivity (h^{-1})	Pressure (bar)	Specific productivity (h^{-1})
Acetone to isopropanol	50	0.4	10	3.5
Furfurol to tetrahydrofurfuryl alcohol	150	0.1	50	0.25
4-nitrosophenol to 4-aminophenol (solvent–ethanol)	50	0.2	50	0.3
Furfurol to furfuryl alcohol	100–150	0.15	50	0.45
Nitroparaffins (C12–C14) to aminoparaffins (solvent–methanol)	50	0.15	50	0.45
Dinitrotriethylbenzene to diaminotriethylbenzene (solvent–methanol)	50	0.15	50	0.3
Dinitrotriethylbenzene to diaminotriethylbenzene (without solvent)	50	∼0	50	0.15
1,5-dinitronaphtalene to 1,5-diaminonaphtalene (10 % suspension)	This technology is not suitable for the reaction		50	0.1
Nitrobenzene to aniline	50	0.15	50	0.4
3,4-diclornitrobenzine to 3,4-dicloraniline (solvent–toluene)	200	0.2	50	0.4
2,4/2,6-dinitotoluene to 2,4/2,6-diaminotoluene	50	0.15	50	0.4
1-octene to octane (1.0 mm particles)	25	5	25	32
1-octene to octane (0.6 mm particles)	This technology is not suitable for such small particles		25	65
1-octene to octane (0.35 mm particles)	This technology is not suitable for such small particles		25	500

Energy consumption for liquid recirculation in the POLF technology is 100–1500 times less than that for gas recycling in TBR and BCR. Catalyst and temperature at reactor inlet are the same for each given reaction (after [6])

9.2.6 Comparison of the POLF Technology with the Conventional Fixed-Bed Reactors

The comparison of the one-stage POLF technology with the conventional TBRs and BCRs processes is shown in Table 9.2.

As can be seen, the POLF technology is characterized by incredibly high productivity under far less pressure. Putting together Tables 9.1 and 9.2, one can also see that the POLF technique is undoubtedly more preferable than the GIPKh processes.

Fig. 9.5 The MPCP GmbH
pilot plant for deep
hydrodesulphurization of jet
fuel. The adjustable reactor
position is designated for
simulations on board
(reprinted from [7] with
permission from Elsevier)

Apart from the extremely high productivity and, therefore, compact design, some other advantages of the POLF technology compared to other techniques are worthy of note.

(i) Energy consumption for liquid recirculation lies in the same range as in the GIPKh processes, i.e., about 100 and more times less than the energy demanded for gas loop in the conventional fixed-bed reactors.

(ii) Obviously, the POLF technology is the only one that can utilize small catalyst particles in industrial multiphase fixed-bed reactions. Table 9.2. indicates that a decrease in the catalyst size (see hydrogenation of 1-octene—0.35 or 0.6 mm vs. 1.0 mm) leads to the further process intensification. Some industrial units in bulk and fine chemistry, if they are designed according to the POLF technology, will be as small as a pilot-plant installation.

(iii) The POLF technology can process the bad soluble compounds, the liquid feed of which represents a suspension of solid as, for instance, in hydrogenation of 1,5-dinitronaphtalene (Table 9.2).

(iv) The POLF system is the safest among all other techniques. Because of the low concentration of the dissolved gas and its insignificant amount in the whole system, "hot spots" as well as the temperature excursion or runaway are never possible.

(v) An insignificant temperature rise along the reactor and complete wetting of a catalyst results in the high selectivity and longer catalyst lifecycle.

(vi) The POLF technique is also more appropriate for purification processes, for example, for ultra-deep hydrodesulphurization. In hydrodesulphurization of 4, 6-dimethyl-dibenzothiophene, the POLF reactor demonstrates about four times higher productivity [12].

(vii) Because the POLF reactor is operable in any position, it can be used on board. The pilot-plant unit for ultra-deep hydrodesulphurization of jet fuels is shown in Fig. 9.5. At liquid hourly space velocity of about 0.7 h^{-1} and partial hydrogen pressure of 20 bar, the POLF reactor removes the sulfur compounds down to 1 ppm [7].

9.3 Concluding Remarks

The unusual scientific principles realized in the GIPKh and, especially, POLF technologies not only challenge the traditional, deep-rooted ideas in multiphase catalysis (Chaps. 3–7), but also demonstrate the significant importance in industrial practice.

Both these techniques allow for the safe operation under the pressure, which is several times less than that in the traditional TBRs and BCRs. Despite the far less operating pressure, the concentration of the reacting gas on the catalyst surface in these processes (especially in the POLF reactors) is much higher than in TBRs and BCRs. As a result, the POLF reactors exhibit much higher catalyst productivity (tenfold and more) as well as better process selectivity and dynamics of the catalyst aging. Moreover, POLF reactors can utilize small catalyst particles (less than 1 mm) without any problem to pressure drop. It is important to point out that the efficiency of the energy consumption in these reactors is much higher than in the traditional TBRs and BCRs. and, therefore, operating costs are significantly less.

References

1. L.B. Datsevich, D.A. Muhkortov, Multiphase fixed-bed technologies. Comparative analysis of industrial processes (Experience of development and industrial implementation). Appl. Catal. A **261**(2), 143–161 (2004)
2. L. Datsevich, M. Nagrodskii, G. Ryleev, G. Tereshenko, Y. Sharikov, *Continuous process for liquid-phase catalytic hydrogenation of organic compounds (in Russian)*, SU Patent 146092,: 1988
3. I. Bat', A. Burtsev, L. Datsevich, G. Mironova, M. Nagrodskii, P. Ovchinikov, G. Ryleev, Y. Sharikov, G. Tereshenko, *Process for production 3,4-dichloraniline (in Russian)*, SU Patent 1392845, 1988
4. L.B Datsevich, I. Golubkov, A. Grachev, L. Grankina, Y. Grigor'ev, M. Kambur, O. Kuznetsova, M. Nagrodskii, G. Ryleev, O. Sokolova, *Process for production of tetrahydrofurfuryl alcohol (in Russian)*, SU Patent 1460944, 1988

5. L. Datsevich, D. Mukhortov, *Process for hydrogenation of organic compounds*, RU Patent 2083540, 1997
6. L.B. Datsevich, D.A. Mukhortov, Saturation in multiphase fixed-bed reactors as a method for process intensification/reactor minimization. Catal. Today **120**, 71–77 (2007)
7. L.B. Datsevich, F. Grosch, R. Köster, J. Latz, J. Pasel, R. Peters, T. Pohle, H. Schiml, W. Wache, R. Wolfrum, Deep desulfurization of petroleum streams: Novel technologies and approaches to construction of new plants and upgrading existing facilities. Chem. Eng. J. **154**, 302–306 (2009)
8. A. Jess, L. Datsevich, N. Gudde, *Purification process*, WO Patent 03091363, 2003
9. L.B. Datsevich, M.P. Kambur, D.A. Mukhortov, Hydrotreating in processes of recuperation of spent oils from motors and electric transformers, *Fuels and Lubricants (in Russian)* **2001**, *3* (on line publication http://www.apris.ru/?page=static&section=51)
10. L.B. Datsevich, F. Grosch, R. Wolfrum, Zentrifugalpumpe, DE Patent application 102006044579, 2008
11. T.K. Scherwood, R.L. Pigford, C.R. Wilke, *Mass transfer* (Mass Transfer; McGraw-Hill, New York, NY, 1975)
12. C. Schmitz, L. Datsevich, A. Jess, Deep desulfurization of diesel oil: kinetic studies and process-improvement by the use of a two-phase reactor with pre-saturator. Chem. Eng. Sci. **59**, 2821–2829 (2004)

Chapter 10
Conclusions and Perspectives

In this book, the author has tried to show that the traditional ideas in design of three-phase fixed-bed reactors are erroneous. Many of these ideas represent nothing more than myths and misconceptions, which, nevertheless, are widely spread among academics and engineers.

The analysis in this monograph clearly proves that the conventional multiphase fixed-bed technologies with gas recirculation do not possess any potential for the process enhancement even if active catalysts are used. The thermodynamic restrictions imposed on the gas–liquid flow as well as the extremely low efficiency of energy utilization makes any attempt of the process improvement in terms of Fig. 1.2 useless.

As is shown, the POLF reactors conflict with the established ideas, but demonstrate superiority over the traditional technologies. This again refutes the myths and misapprehensions attributed to the conventional techniques.

The author believes that the critical analysis of the chemical and physical processes presented in this monograph may initiate a reconsideration of the traditional scientific and technological paradigms and give a fresh impetus to new technological approaches in design of industrial reactors.

From the author's point of view, the future steps in industrial development should be concentrated on the further implementation of the principles lying in the POLF technique and the oscillation theory:

(1) Development of the multistage "oversaturated" POLF process. This technique can be useful for a variety of industrial applications, e.g. in oxidation, hydrogenation and the Fischer–Tropsch synthesis. It is characterized not only by the constant concentration of the gas compound in the liquid phase, but also by the plug-flow pattern, which results in very much higher catalyst productivity than shown in Table 9.2. In addition, an industrial plant can be designed as a diminutive system of high integrity (by analogy to a microchip).

(2) Induction of liquid oscillations and physical pumping in catalyst pores [1, 2]. The noticeable enhancement of the reaction rate should be expected.

L. B. Datsevich, *Conventional Three-Phase Fixed-Bed Technologies,*   97
SpringerBriefs in Applied Sciences and Technology,
DOI: 10.1007/978-1-4614-4836-5_10, © The Author(s) 2012

References

1. B. Blümich, L.B. Datsevich, A. Jess, T. Oehmichen, X. Ren, S. Stapf, Chaos in catalyst pores: can we use it for process development? Chem. Eng. J. **134**, 35–44 (2007)
2. Movie 5—Catalyst engineering: induced pumping through a catalyst particle in the reaction of hydrogen peroxide decomposition, MPCP GmbH, Illustrative material to the oscillation theory. http://mpcp.de/en/research_and_development/oscillation_model/illustrative_material/

Appendix A
Evaluation of an Incorporated Heat Exchanger Destined for the Complete Heat Withdrawal from a Catalyst Bed

In order to show the hopelessness of any attempt to remove the reaction heat only by heat exchangers combined with the catalyst bed, we consider the optimistic variant, which is characterized by the low reaction rates, low heat production, and comparatively high heat transfer properties of a hypothetic heat exchanger. As an example, we take the hydrogenation reaction of 1-hexene to n-hexane, the heat effect of which is relatively low if compared to other reactions enumerated in Table 1.1.

The reaction rate for this reaction is taken from experiments under the conditions when a catalyst particle (6 × 6 mm) was completely submerged in the liquid phase without any stirring [1]. It can be expected that the reaction rate in an industrial reactor (if somebody incurs a risk of its operation only with an embedded heat exchanger), should be significantly higher by virtue of more intensive gas–liquid–solid mass transfer.

The reaction parameters and heat transfer characteristics used in this evaluation are given in Table A.1.

It is necessary to point out once more that not only the reaction rate, but also other features in Table A.1 represent the best case with regard to a more compact heat exchanger than it can be expected in reality. For instance, the overall heat transfer coefficient is too upbeat. Actually, the catalyst bed possesses the extremely bad properties for heat transfer, which may result in an inadmissible temperature gradient if the space between heat transfer interfaces is more than several centimeters (wall-side heat transfer coefficients for TBR can be found, for example, in [2]).

In the hypothetic reactor, let us consider the catalyst bed just at the entrance of the liquid compound that is pierced by the heat exchanger tubes, the diameter of which is d_{HE}.

The overall heat production with respect to a volume of the catalyst bulk V can be calculated as

$$Q_{HE} = V(1 - \varepsilon)r_V(-\Delta H) \tag{A.1}$$

L. B. Datsevich, *Conventional Three-Phase Fixed-Bed Technologies,*
SpringerBriefs in Applied Sciences and Technology,
DOI: 10.1007/978-1-4614-4836-5, © The Author(s) 2012

Table A.1 Data for heat transfer evaluation

Reacting system		
1	Reaction	$C_6H_{12} + H_2 \rightarrow C_6H_{14}$
2	Catalyst	Ni catalyst 6 × 6 mm (NISAT Sued-Chemie)
3	Bed porosity of catalyst (ε)	0.53
4	Heat of reaction related to a mole of reactant ($-\Delta H$), J/mol	125×10^3
Operating conditions and reaction rate		
1	Temperature, °C	160
2	Total pressure in the reactor (P), bar	101
3	Partial pressure of hydrogen (P_B), bar	90
4	Concentration of hydrogen in the liquid bulk ($C_{B,l}$) mol/m^3	760
5	Concentration of 1-hexene in the liquid bulk ($C_{A,l}$), mol/m^3 (% mass)	5,990 (96 % mass)
6	Observed reaction rate related to a mole of n-hexane per volume of the catalyst particle (r_V), mol/(m^3 × s)	475
Heat transfer properties for a hypothetic heat exchanger		
1	Heat transfer coefficient (α), W/(m$^2 \cdot$ K) (kcal/(m$^2 \cdot$ K $\cdot$ h))	1,160 (1,000)
2	Log mean temperature difference (ΔT_{HE}), K	50

Table A.2 Specifications of the hypothetic heat exchanger related to 1 m^3 of the catalyst bulk

Tube diameter d_{HE} (cm)	Total length of tubes[a] L_{HE} (m)	Volume of heat exchanger tubes[b] V_{HE} (m^3)
1	15,300	1.2
2	7,700	2.4
3	5,100	3.6
4	3,100	6.0
5	4,000	7.8

[a] $L_{HE} = F_{HE}/(\pi d_{HE})$ [b] $V_{HE} = F_{HE}d_{HE}/4$

The same amount of heat should be removed through the heat transfer surface according to

$$Q_{HE} = \alpha F_{HE}\Delta T_{HE} \tag{A.2}$$

From Eqs. (A.1) and (A.2), one can yield the ratio of the heat transfer surface to the catalyst bed volume as

$$\frac{F_{HE}}{V} = \frac{(1 - \varepsilon)r_V(-\Delta H)}{\alpha \Delta T_{HE}} = 481 \; \frac{m^2}{m^3_{cat}} \tag{A.3}$$

Table A.2 specifies the main features of the heat exchanger related to 1 m^3 of the catalyst bulk.

For example, if the conventional heat exchanger tubes of 2 cm are applied in the catalyst cooling, their total length should be about 7 km with the volume of 2.4 m^3.

Appendix B
Energy Demand for the Compression and Transportation of Gas

Compressing in a single casing compressor can be regarded as an adiabatic process. The power E demanded for the adiabatic compression of the recycled gas with molar flow rate $N_{g,\text{recycle}}$ from pressure P_1 to $P_1 + \Delta P_\Sigma$ can be calculated as

$$E = N_{g,\text{recycle}} C_{V,g} T_1 \left[\left(\frac{P_1 + \Delta P_\Sigma}{P_1} \right)^{\frac{\gamma-1}{\gamma}} - 1 \right] \tag{B.1}$$

where T_1 and P_1 being the temperature and pressure of gas at suction, ΔP_Σ being the pressure rise developed by compressor, $C_{V,g}$ being the specific heat for constant volume, and $\gamma = \dfrac{C_{P,g}}{C_{V,g}}$ being the adiabatic index of recycled gas.

If $\dfrac{\Delta P_\Sigma}{P_1} \ll 1$, Eq. (B.1) can be transformed to

$$E \approx N_{g,\text{recycle}} \frac{R}{\gamma} T_1 \frac{\Delta P_\Sigma}{P_1} \tag{B.2}$$

For hydrogen, for example, Eq. (B.2) produces the accuracy of below 15 % compared to Eq. (B.1) when $\dfrac{\Delta P_\Sigma}{P_1} < 0.45$. Since this condition is true for many industrial processes, it can be concluded that the energy demanded by the recycle compressor for gas recirculation is proportional to the temperature T_1 and the pressure rise ΔP_Σ and inversely proportional to the pressure at the suction P_1.

If the transportation of recycled gas through any element of a gas loop is regarded as isothermal, the energy dissipation provided by gas flowing at temperature T from the initial pressure P_1 to $P_1 - \Delta P$ can be written as

$$E = N_{g,\text{recycle}} RT \ln \left(\frac{P_1}{P_1 - \Delta P} \right) \tag{B.3}$$

L. B. Datsevich, *Conventional Three-Phase Fixed-Bed Technologies*, 103
SpringerBriefs in Applied Sciences and Technology,
DOI: 10.1007/978-1-4614-4836-5, © The Author(s) 2012

where P_1 being the pressure at the inlet of the considered element and ΔP being the pressure drop over the considered element.

If $\dfrac{\Delta P}{P_1} \ll 1$, one can obtain from Eq. (B.3)

$$E \approx N_{g,\text{recycle}} RT \frac{\Delta P}{P_1} \tag{B.4}$$

Equation (B.4) can be used with the accuracy of below 15 % when $\dfrac{\Delta P}{P_1} < 0.3$.

As is seen, the energy dissipated at gas transportation is proportional to the temperature T and the pressure drop ΔP and inversely proportional to the initial pressure P_1.

References

1. T. Oehmichen, Einfluss der Gas/Dampfblasenbildung auf die effektive Kinetik heterogen-katalysierter Gas/Flüssig-Reaktionen (Ph.D. Thesis). Schaker Verlag, Aachen, 2010
2. V. Specchia, G. Baldi, Heat transfer in trickle-bed reactors. Chem. Eng. Commun. **3**, 483–499 (1979)

Index

A
Adiabatic reactor, 70
Adiabatic temperature, 20, 35, 56, 63,
 70, 71, 78
Admissible temperature, 34, 78, 99
Adsorption, 14, 15, 45
Aging, 13, 16, 23, 33, 34, 45, 58, 82,
 89, 91, 92
Attrition, 8

B
Balance, 1, 34, 35, 56, 78, 79, 90
Bubbles, 39, 40, 42, 47, 88, 89
Bubble (packed)
 Column Reactors, 1, 6
BCR, 1, 23, 25, 27, 28, 34–36, 43, 46–48, 57,
 58, 61, 63, 71, 74, 80, 82–85, 89,
 90–92, 94

C
Capillary, 12, 40, 43
Coke, 13, 16, 36
Column Reactors
Compressor, 6, 22, 23, 25–31, 36, 38, 39,
 45, 53–57, 59, 65, 66, 70, 73,
 75–77, 103
Conversion, 13, 19, 35, 48, 51, 59, 71, 76–78,
 80, 82, 88, 89
Coolant, 76
Concurrent, 25, 34, 75, 76, 81
Countercurrent, 30, 75, 76, 81
Cracking, 4, 6, 21, 24, 30, 42, 51, 60, 62
Crush strength, 26, 36

D
Decay, 13, 46
Deoxidation, 4
Deposition, 45
Diffusion, 9, 14, 41, 52, 91
Diffusivity, 39, 49, 57
Design, 4, 5, 11–14, 16, 19, 20, 22–24, 47,
 58–60, 64, 66, 77, 81, 82, 84, 88, 91,
 92, 97
Desorption, 14, 45
Distribution, 12, 23, 28, 43, 54, 58,
 70, 79

E
Effectiveness factor, 12, 66
Enhancement, 39, 66, 69, 84, 90, 91, 97
ER (Eley-Rideal), 14
Equilibrium, 17, 18, 47, 63, 65,
 86–89, 91
Equipment, 6, 22, 23, 30, 60, 73, 74, 77, 85,
 89, 91
Evaporation, 35, 42, 55, 56, 81, 82
Excess, 16, 48, 63, 89
Excursion, 56–58, 70, 84, 93
Exothermic reaction, 8, 34, 40, 42, 47, 61, 69,
 70, 71, 82
External mass transfer, 14, 17, 46, 47,
 48, 65, 85

F
Film, 43, 47, 70, 73, 75, 80–82, 85
Filtration, 7, 8, 42
Fischer-Tropsch synthesis, 1, 3, 97

L. B. Datsevich, *Conventional Three-Phase Fixed-Bed Technologies*,
SpringerBriefs in Applied Sciences and Technology,
DOI: 10.1007/978-1-4614-4836-5, © The Author(s) 2012